Study & Master

PHYSICAL SCIENCE
GRADE 10

D1824579

E.J. van Zyl
V. Craül
A. Meyer
E. Oosthuizen

ROEDURICO
t r u s t

Somerset West

PUBLISHED BY THE PRESS SYNDICATE OF
THE UNIVERSITY OF CAMBRIDGE
The Pitt Building, Trumpington Street, Cambridge, United Kingdom

CAMBRIDGE UNIVERSITY PRESS
The Edinburgh Building, Cambridge CB2 2RU, UK
40 West 20th Street, New York, NY 10011–4211, USA
477 Williamstown Road, Port Melbourne, VIC 3207, Australia
Ruiz de Alarcón 13, 28014 Madrid, Spain
Dock House, The Waterfront, Cape Town 8001, South Africa

http://www.cambridge.org

First published 2001
Second (revised) impression 2002
Third impression 2002
Fourth impression 2003

Printed by ABC Press, Kinghall Avenue, Epping, South Africa

Typeset in 10 on 11 pt Times Roman
Illustrations by Arno Burger and Sunita Joyce

A catalogue record for this book is available from the British Library

ISBN 0 947465 68 5   paperback

Preface

To the Educator

Study and Master Physical Science 10 is written to expand and consolidate the basic Physical Science knowledge required by the syllabus. It will provide assistance with the continuous assessment and examining of learners. It provides a source of information for **class work, homework** and **revision.** The questions can be used for cumulative assessment and examining of learners.

In the modern world, where Physical Science is a fundamental building block of Education, it is of major importance that learners remain interested in Physical Science, hopefully until the end of their school career. It is as a result of this that the solutions which enhance the **learners' reference framework** at home during self-study, appear at the end of the chapter. This textbook is also useful where there are large classes and where learners do not have access to additional Physical Science instruction.

With a little initiative and adaption of homework given to learners, this textbook could be of cardinal importance to all learners. The logical compilation of the book is thus as follows:
• Subject-matter is presented in a concise, straightforward and factual manner.
• Main sections are divided into shorter sub-sections to facilitate learning, reference and revision.
• Formulation and application of scientific facts and principles.
• Appropriate exercises with calculated solutions.
• Multiple choice, objective and structured questions with answers.
• Subject-matter which applies to the **Higher Grade** only is clearly indicated.

Higher Grade learners should work through **all** the sections of the book. The **same** applies to **Standard Grade learners** with the exception of those sections indicated for Higher Grade only.

To the learner

This is a learner-friendly book written specially for you to help you come to grips with the requirements of the syllabus. It is important that you use this book for **independent self-study.**

Physical Science is a key-subject in the school curriculum and if you use this book in the correct disciplined way, you will surely reap the benefits. You will also, simultaneously, develop a greater sense of responsibility, more self-discipline as well as greater self-motivation.

• **Read** the relevant sections in the chapter, that has been dealt with, attentively.
• Summarize, in **point-note form**, the subject-matter that was dealt with in class.
• Constantly ensure that you understand what the examples are explaining to you.
• **Anwer** the **questions** at the end of each chapter. Under no circumstances should you read the answers before working through the **questions yourself first**.
• **Evaluate** your own work by comparing your answers with those in the book.
• Should **problems** arise, consult your Educator.
• Use this textbook independently and remember that your Educator is going to make certain adaptations in the tests and examinations; that is why it is important to constantly understand **what** should be done and **how** it should be done.

Best of luck! We hope this book will bring about positive change.
The Authors

Study & Master series

Grade 8

General Science
Mathematics

Grade 9

General Science
Mathematics

Grade 10

Biology
Mathematics
Physical Science

Grade 11

Biology

Grade 12

Biology

Grade 11 and 12 (one book)

Accounting
Agricultural Science
Mathematics Higher Grade
Mathematics Standard Grade
Physical Science

2002   **Matric Question Papers (one book)**

8 in ONE: Accounting; Biology; Physical Science and Mathematics
(SG and HG) *2001 Senior Certificate Examination Questions papers and memorandums*

Curriculum 2005

Grade 4 – Stepping into Natural Sciences and Technology
Grade 5 – Stepping into Natural Sciences and Technology
Grade 6 – Stepping into Natural Sciences and Technology
Grade 7 – Journeying into Natural Sciences
Grade 8 – Journeying into Natural Sciences
Grade 9 – Journeying into Natural Sciences

Tips for Teens: Life Orientation

Survive your Life
Survive your Studies

Roedurico Trust
Tel: (021) 852-3104
Fax: (021) 851-6874

Contents

1 *Waves*

1. Vibrations

A vibration is a regular repetitive movement. It means that the movement repeats itself and each complete repetition takes the same time. It can be a back-and-forth, up-and-down or a rotation movement.

1.1 Terminology

Consider the movement of a swinging pendulum.

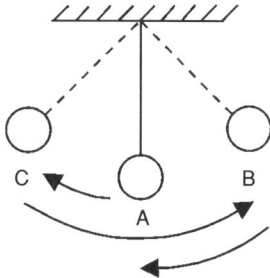

- Point A is the point of equilibrium or rest.

- Points B and C are the points of maximum displacement from the equilibrium (point A).

- The amplitude of the vibration is the magnitude of the maximum displacement from the equilibrium position.

- The movement from point A to point B, then to point C, and back to A, is called one vibration cycle.

1.2 Period and frequency

The period (T) of a vibration is the time required to complete one full vibration. Periods are measured in seconds (s).

Therefore: $\text{period} = \dfrac{\text{time lapse}}{\text{number of vibration cycles}}$

The frequency (f) of a vibration is the number of vibration cycles completed per second. Frequency is measured in hertz (Hz).

Therefore: $\text{frequency} = \dfrac{\text{number of vibration cycles}}{\text{time lapse}}$

The relationship between frequency (f) and period (T) is represented by $f = \dfrac{1}{T}$ or $T = \dfrac{1}{f}$.

Example:

1. In an experiment it is established that a swinging pendulum completes 12 full swings in one minute.

Calculate

(a) the frequency of the movement;
(b) the period of the movement.

---

2. Waves

2.1 Pulses

A pulse can be considered as a disturbance that is conducted through a medium, through which energy is transferred from one point to another.

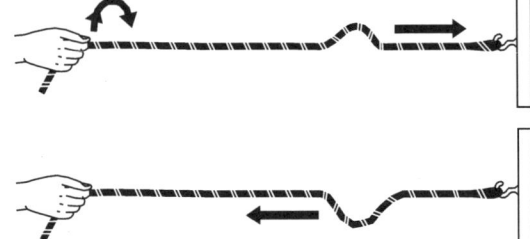

For example, a pulse can arise when one point of a spiral spring is given a tug so that a disturbance moves down the spring.

2.2 Waves

All waves arise due to vibrations or oscillations.

A wave is a means by which energy is transferred from one point in a medium to another by a series of consecutive pulses.

Two types of waves exist, namely **transverse waves** and **longitudinal waves**.

2.3 Transverse Waves

A **transverse wave** can be formed by securing one point of a spiral spring and moving the other point **back-and-forth**. The direction in which the medium (spring) moves is **perpendicular** to the direction that the wave moves.

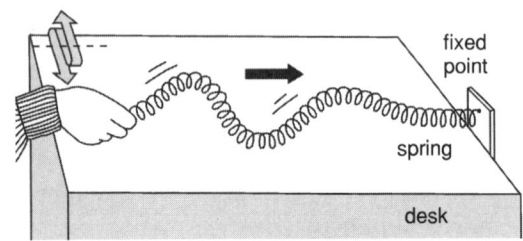

A transverse wave is obtained when the disturbance of the medium is perpendicular to the direction of propagation of the wave.

Examples of transverse waves are water waves, light waves and radio waves.

Terms

- The **frequency (f)** is the number of wave pulses per time unit that move past a fixed point.

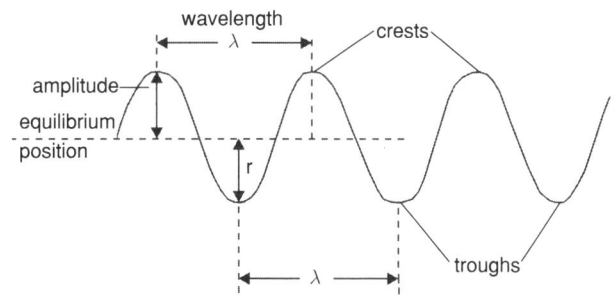

- The highest parts of the wave are called the **crests** and the lowest parts the **troughs**.

- Two points move **in phase** if they are performing the same movement at the same time (e.g. two crests would be in phase as well as two troughs).

- The **wavelength (λ)** is the distance between two consecutive points that are in phase (e.g. two consecutive crests), and is measured in metres (m).

- The **amplitude (r)** is the maximum displacement of the wave from its equilibrium position, measured in metre (m).

2.4 Longitudinal Waves

A **Longitudinal wave** can be formed by securing one point of a spiral spring and moving the other point **forward-and-back**, so that a series of compressions and rarefactions are formed. The direction in which the spring vibrates is **parallel** to the direction in which the wave is moving.

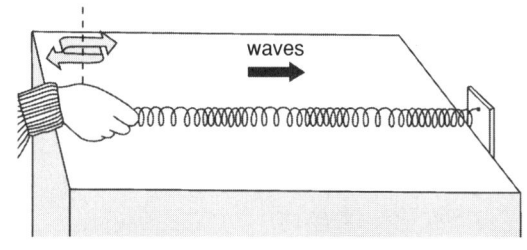

A longitudinal wave is obtained when the disturbance of the medium is parallel to the direction of propagation of the wave.

An example of longitudinal waves is sound waves.

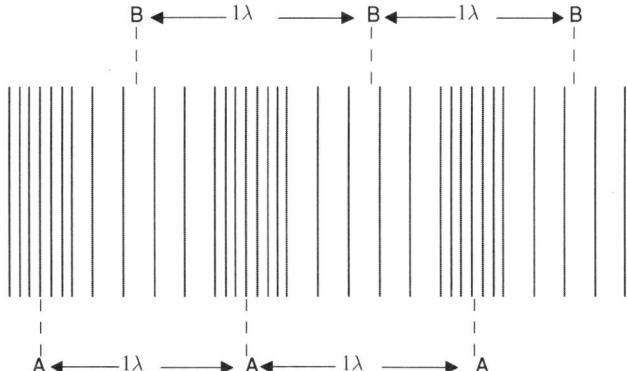

Terms

- One wavelength is the distance between two consecutive compressions (points A) or two consecutive rarefactions (points B).

3. Speed of Waves

3.1 The relationship between speed, wavelength and frequency of a wave

The speed by which a wave is conducted is the distance that a wave pulse completes per second.

Therefore: $\text{speed} = \dfrac{\text{distance (m)}}{\text{time (s)}}$

The unit in which speed is measured is m.s⁻¹.

The relationship between speed (v), wavelength (λ) and frequency (f) of a wave is given by the general wave equation that applies to all waves:

$$\textbf{Speed} = \textbf{frequency} \times \textbf{wavelength}$$

Or in symbols: $v = f \times \lambda$

There are two formulas with which the speed of a wave can be calculated:

$\text{speed} = \dfrac{\text{distance}}{\text{time}}$ and $\text{speed} = \text{frequency} \times \text{wavelength}$.

Examples 1:

The speed of propagation of a certain wave is 8 m.s⁻¹. If the wavelength is 400 mm, calculate the frequency of the wave.

Solution:

speed (v) = 8 m.s⁻¹
wavelength (λ) = 400 mm = 0,4 m
$v = f\lambda$
$\therefore 8 = f \times 0,4$
$\therefore f = \dfrac{8}{0,4} = 20 \text{ Hz}$

Example 2:

The diagram represents the profile of a wave series with a frequency of 20 Hz.

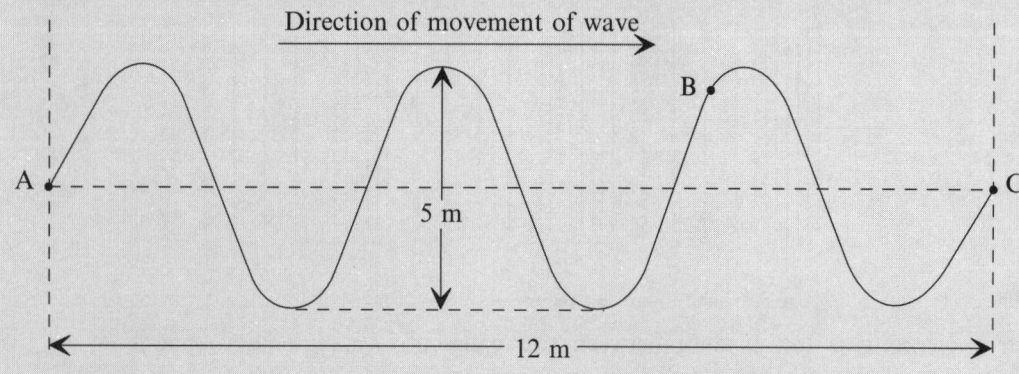

(a) How long does it take for three waves to move past point C?
(b) Calculate the wavelength of the waves.
(c) What is the amplitude of the waves?
(d) In which direction is point B moving?
(e) Calculate the speed of the waves in two ways.

Solution:

(a) $f = 20$ Hz

$\therefore T = \dfrac{1}{f} = \dfrac{1}{20} = 0,05$ s

Time lapse for one wave = $0,05$ s

$\therefore$ Time lapse for 3 waves = $0,05 \times 3 = 0,15$ s

(b) length of 3 waves (A → C) = 12 m

$\therefore \lambda = \dfrac{12}{3} = 4$ m

(c) $A = \dfrac{5}{2} = 2,5$ m

(d) downwards (the wave is already past point B)

OR:

Draw the wave movement at B a short moment later. Draw a vertical line through point B and see where the "new" point B lies.

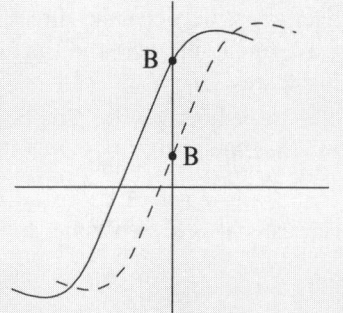

(e) (i) $v = f\lambda$

$\quad = 20 \times 4 = 80$ m . s^{-1}

(ii) speed $= \dfrac{\text{distance}}{\text{time}} = \dfrac{12}{0,15} = 80$ m . s^{-1}

Example 3:

A boy on the harbour wall watches the sea waves that move past and observes that a wave goes past every 8 seconds. The distance between 4 consecutive wave crests is 45 m. Calculate the wave's
(a) frequency.
(b) wavelength.
(c) speed.

Solution:

(a) $f = \dfrac{\text{vibrations}}{\text{time}} = \dfrac{1}{8} = 0,125$ Hz

(b) The distance between 4 wave crests is 3 waves

$\therefore \lambda = \dfrac{45}{3} = 15$ m

(c) $v = f\lambda = 0,125 \times 15 = 1,88$ m . s^{-1}

OR:

speed $= \dfrac{\text{distance}}{\text{time}} = \dfrac{15}{8} = 1.88$ m . s^{-1}

3.2 Reflection of waves

Terms

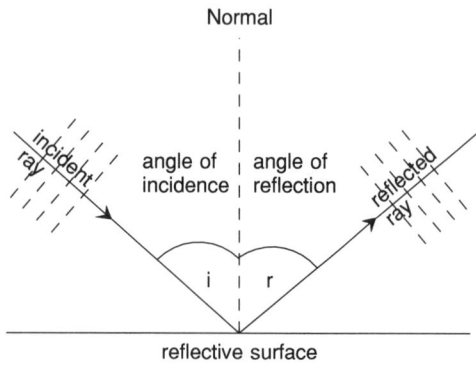

- **Shallow waves,** such as those obtained in a wave tank, cause straight, parallel wave fronts.
- The **normal** is a line that lies perpendicular to the reflective surface.
- The direction in which the wave moves, is indicated with arrows that are called **rays**.
- The angle between the direction of the incoming wave (incoming ray) and the normal is the **angle of incidence ($\angle$i).**
- The angle between the direction of the reflected wave (reflected ray) and the normal is the **angle of reflection ($\angle$r).**

When shallow waves strike an obstruction, they are reflected in such a way that the **angle of incidence ($\angle$i) is always equal to the angle of reflection ($\angle$r).**

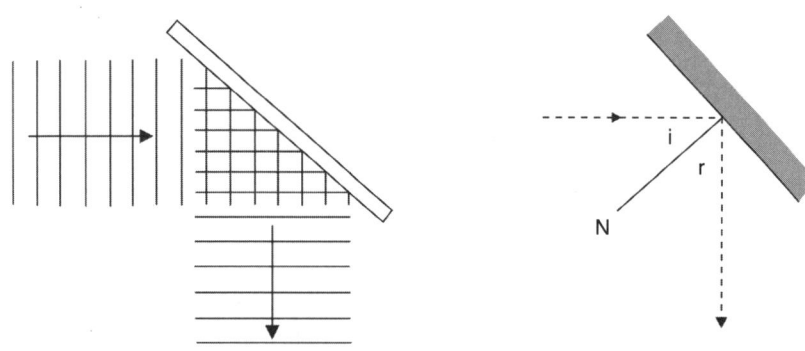

A **convex** obstruction causes the shallow waves to spread out, or **diverge**, when they are reflected.

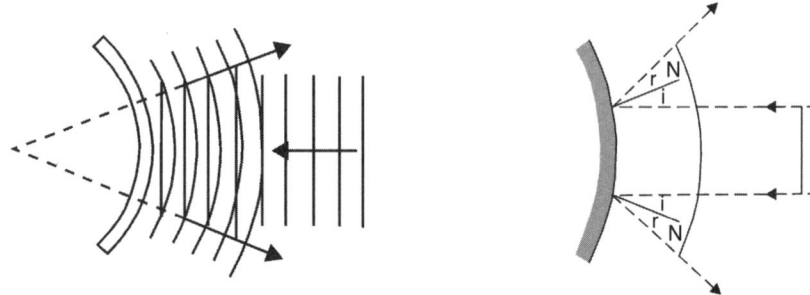

A **concave** obstruction causes the shallow waves to come together at a point, or **converge**, when they are reflected.

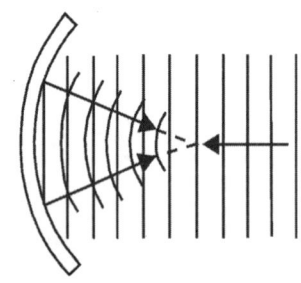
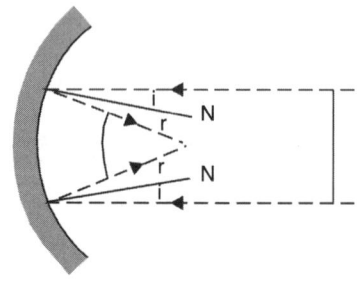

6

3.3 Refraction of waves

When the shallow waves in a wave tank reach an area of shallower water perpendicularly, the speed of the wave decreases. Seeing that the frequency of the wave remains the same, the wavelength of the wave becomes shorter when it enters the area of shallower water.

> When the waves strike the dividing surface between deep and shallower water obliquely, the decrease in speed causes the waves' direction to change. **Refraction** has taken place.

The **refracted ray** indicates the direction of the wave's motion after refraction has taken place.

The **angle of refraction ($\angle$ r)** is the angle between the refracted ray and the normal.

When the waves move from **deeper to shallower** water and strike the dividing surface obliquely, the wave's direction is bent towards the normal. The angle of incidence is **greater** than the angle of refraction ($\angle i > \angle r$).

When the waves move from **shallower to deeper** water and strike the dividing surface, the waves' direction is bent away from the normal. The angle of incidence is **smaller** than the angle of refraction ($\angle i < \angle r$).

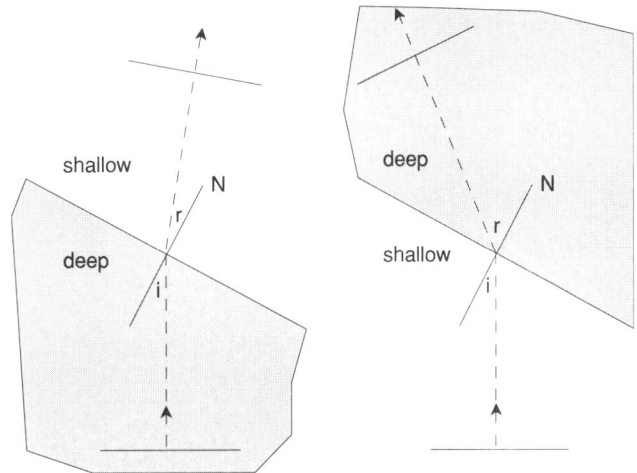

3.4 Interference of waves

When two wave pulses approach each other and cross, **interference** takes place.

When two wave crests of identical wave pulses meet, **constructive interference** takes place. The crests reinforce each other and form a wave pulse with a greater amplitude.

When a crest and a trough of identical wave pulses meet each other, **destructive interference** takes place. The crest and the trough cancel each other out and no movement takes place.

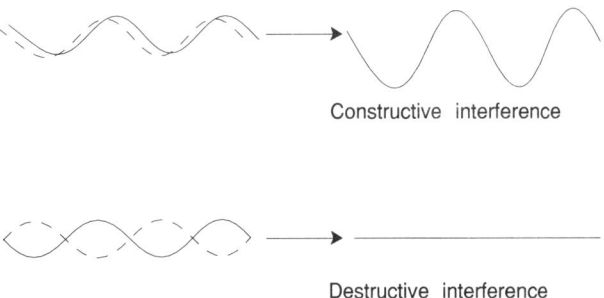

Constructive interference

Destructive interference

> **Constructive interference** therefore takes place when two waves **in phase** meet and **destructive interference** takes place when two waves **out of phase** meet.

When two sets of circular waves meet each other in a wave tank, an interference pattern is formed, as illustrated in the sketch.

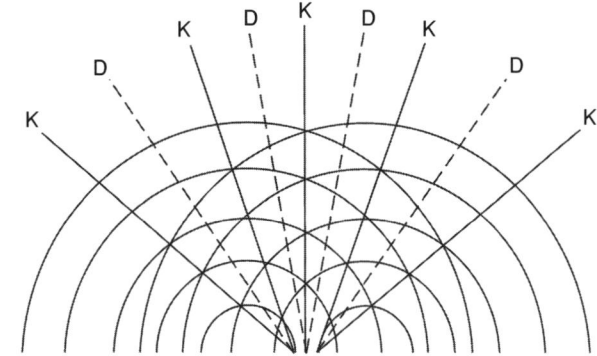

The points where constructive interference takes place, are joined to form lines marked "K", and where destructive interference takes place, they are joined to form lines marked "D".

QUESTIONS

Section A

Different options are proposed as answers to the following questions. Choose the correct answer.

1. Which of the following is not a regular repetitive movement?
   A   A metal ball swinging to-and-fro
   B   A vibrating loudspeaker
   C   The keys of a typewriter, whilst typing normally
   D   The fast to-and-fro movements of the wings of a bee.

2.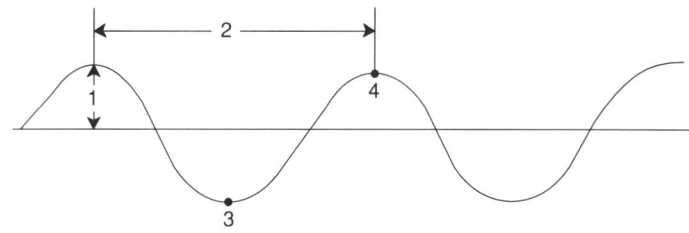

   In the figure the following is represented by 1, 2, 3 and 4:

	1	2	3	4
A	wavelength	amplitude	wave trough	wave crest
B	wave crest	wavelength	wave trough	amplitude
C	wavelength	pulse	amplitude	period
D	amplitude	wavelength	wave trough	wave crest

3. A single disturbance in a medium is called . . .
   A   a wave.   C   the period.
   B   the frequency.   D   a pulse.

4. The frequency of a swinging object is the . . .
   A   number of vibrations that the object completes in one second.
   B   total number of vibrations that the object completes.
   C   time lapse for the object to complete one vibration.
   D   total time lapse for the movement of the object.

5. The distance that any pulse can move in one second, is the wave's . . .
   A   speed.   C   frequency.
   B   period.   D   wavelength.

6. Another unit apart from hertz in which frequency can be measured, is . . .
   A   s.   C s^{-1}.
   B $m.s^{-1}$ D   m.

7. The distance between any two consecutive points in a wave motion that are in phase, is the . . .
   A   period.   C   amplitude.
   B   wavelength.   D   oscillation.

8. The kinetic energy of a water molecule in a water wave is the greatest when it moves through the . . .
   A   trough.   C   equilibrium position.
   B   crest.   D   maximum displacement.

9. Which statement concerning waves is false? Waves . . .
  A   transfer energy.
  B   can be reflected.
  C   always need a medium to be conducted.
  D   can change direction on striking another medium.

10. Only in transverse waves . . .
  A   do the equation $v = f \times \lambda$ apply.
  B   do the particles move across the direction in which the wave is moving.
  C   can interference take place.
  D   can energy be conducted through water.

11. When two waves, moving through the same medium, coincide, they can reinforce or weaken each other. This phenomenon is known as . . .
  A   interference.   C   refraction.
  B   diffraction.   D   reflection.

12. Two wave pulses as shown in the figure, move through a medium and simultaneously reach the same point in the medium.

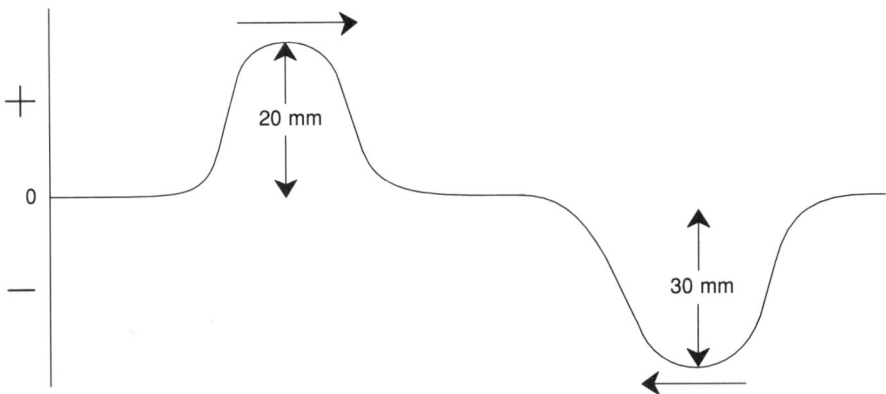

When these two waves coincide, their total amplitude will be . . .
  A   +50 mm   C   −10 mm
  B   +10 mm   D   −50 mm

13. The sketch shows a wave passing from left to right through water.

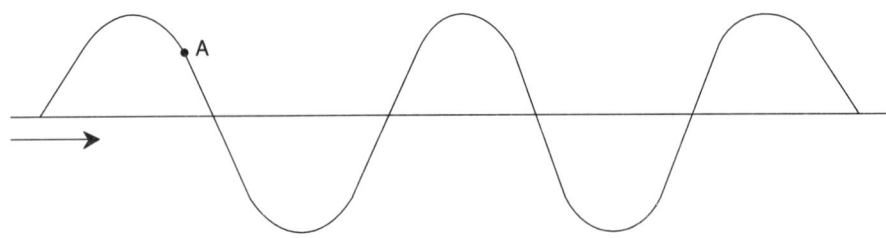

In what direction is a cork at point A moving at this moment?

A   B   C   D

14. The meaning of the symbols in $v = \lambda \times f$ is as follows . . .

	v	**λ**	**f**
A	speed	wavelength	frequency
B	wavelength	amplitude	frequency
C	amplitude	wavelength	speed
D	frequency	speed	amplitude

15. A pendulum completes 20 to-and-fro movements in 5 s. The frequency and period of the pendulum is respectively . . .

	Frequency	**Period**
A	4 Hz	0,25 s
B	20 Hz	5 s
C	4 Hz	20 s
D	0,25 Hz	4 s

16. A wave with a wavelength of 0,15 m moves through water and forms 50 waves in 5 seconds. The speed of the waves is . . .
A   7,5 m . s^{-1}.   C   1,5 m . s^{-1}.
B   0,03 m . s^{-1}.   D   37,5 m . s^{-1}.

17. A wave with a period of 0,2 seconds has a frequency of . . .
A   20 Hz.   C   5 Hz.
B   50 Hz.   D   2 Hz.

18. The pendulum of a clock completes 45 oscillations in one and a half minutes. Its period is . . .
A   30 s.   C   2 s.
B   0,5 s.   D   0,033 s.

Questions 19 to 21 apply to the following diagrammatic representation of a wave motion.

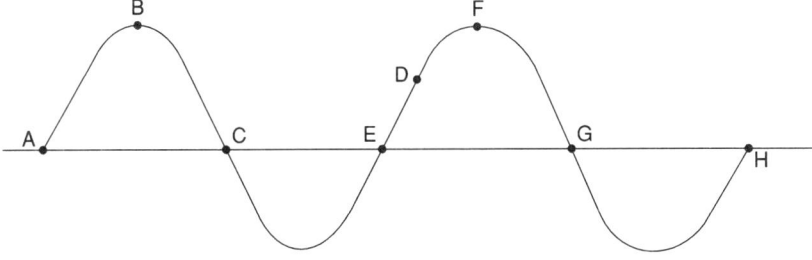

19. Which distance does **not** represent one wavelength?
A   BF   C   AC
B   EH   D   CG

20. Which arrow can represent the direction of energy propagation at D?

A   ↓   B   ↙   C   ↗   D   →

21. If the frequency of the wave is 10 Hz, the time it takes for the wave to transmit energy from point B to point F is . . .

A 0,1 s. C 0,5 s.
B 0,2 s. D 0,05 s.

22. When a wave undergoes a change in speed when passing from one medium to another, the phenomenon is called . . .

A frequency. C reflection.
B interference. D refraction.

23. The phenomenon that a water wave is refracted when it moves between two different depths of water, is due to the fact that the wave's . . .

A speed changes. C wavelength changes.
B frequency changes. D speed, frequency and wavelength change.

24. When a water wave moves from deep to shallow water, its . . .

A frequency decreases. C wavelength decreases.
B speed stays constant. D speed increases.

25. Waves move diagonally from deep to shallow water. If the wavelength of the waves is 10 mm in the deeper water and the angle of incidence of the waves is 30°, possible values for the wavelength and angle of refraction are respectively . . .

A 8 mm and 25°. C 12 mm and 35°.
B 8 mm and 35°. D 10 mm and 25°.

Section B

1.   A vibrating object takes 4 s to complete 20 oscillations.
1.1   Calculate the period of the vibration.
1.2   Calculate the frequency of the vibration.

2.   A pendulum completes 50 to-and-fro movements in 5 s.
2.1   Calculate the
      (a) frequency and
      (b) the period of the pendulum.
2.2   Explain how the
      (a) frequency and
      (b) the period of the pendulum change if the string of the pendulum is lengthened.

3.   Waves in a wave tank have a speed of 0,5 m . s⁻¹ and a wavelength of 25 mm.
3.1   Calculate the frequency of the waves.
3.2   What is the period of these waves?
3.3   When the wave moves to shallower water, the speed decreases to 0,4 m . s⁻¹. Now calculate the wave's
      (a) frequency.
      (b) wavelength.

4. The accompanying sketch is a graphical representation of a transverse wave. Only use the symbols supplied, to indicate the following:

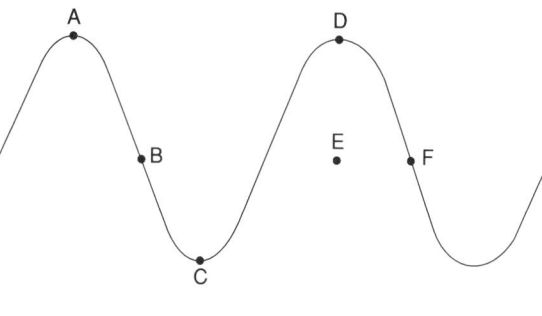

4.1 Any two points that are in the same phase

4.2 Position of equilibrium

4.3 Amplitude

4.4 A trough

4.5 One wavelength

5. The block in the sketch is lifted up to its position at A as shown in the sketch and then released. By **using the given sketch**, explain the following terms:

5.1 Oscillation

5.2 Position of equilibrium (rest)

5.3 A vibration cycle

5.4 Maximum displacement from the position of rest

5.5 Frequency

5.6 Period

5.7 Amplitude

6. Three different waves move at the same speed, but their frequencies and wavelengths differ. Use the information that is given to complete the table.

	speed in m.s^{-1}	frequency in Hz	wavelength in m
Wave 1		8	4
Wave 2		16	
Wave 3			1

7. The wave motion illustrated in the sketch below completes one vibration in 0,5 s.

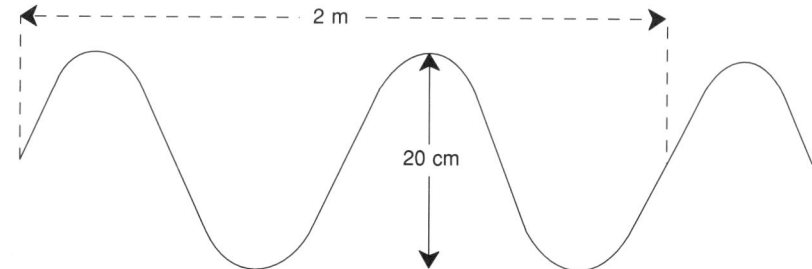

7.1 How long will it take four complete waves to pass any specific point?

7.2 What is the amplitude of this wave motion?

7.3 Calculate the period of the waves.

7.4 Calculate the
(a) wavelength
(b) frequency
(c) speed of the waves.

8. The vibrator in a wave tank generates waves with a frequency of 10 Hz.

8.1 What is meant by the term frequency?

8.2 What type of wave is generated by the vibrator?

8.3 Give the name of another type of wave motion and explain how it differs from the type referred to in the previous question.

8.4 Calculate the wavelengths of the wave motion if the distance between 21 wave crests is 8 mm.

8.5 Calculate the speed of propagation of the waves in $m.s^{-1}$.

9. The diagram illustrates the wave pattern of a wave series with a frequency of 30 Hz.

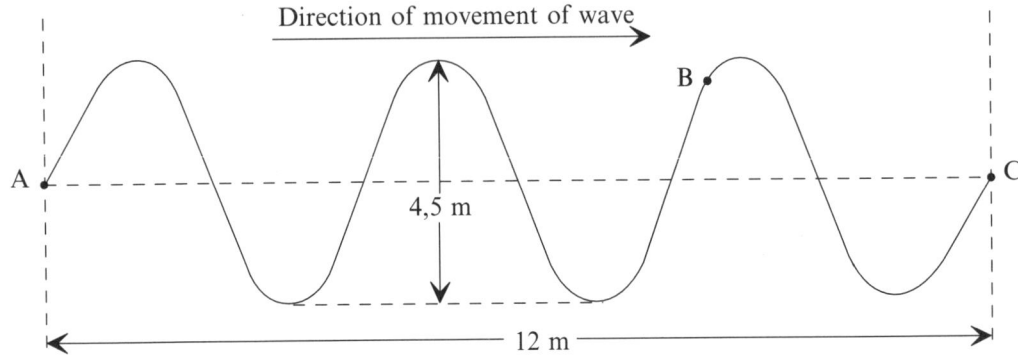

9.1 How much time passed as the waves moved from A to C?

9.2 Calculate the wavelength of the waves.

9.3 What is the amplitude of the wave motion?

9.4 In what direction is B moving at this point?

9.5 Calculate the speed of the waves in two ways.

10. Calculate the wavelength of a wave with a speed of 12 $m.s^{-1}$ and a period of 0,025 s.

11. The distance between 10 consecutive crests in a wave tank is 360 mm. Each crest covers a distance of 400 mm in 2 s.
Calculate

11.1 the propagation speed of the waves.

11.2 the wavelength of the waves.

11.3 the frequency of the waves.

12. In an experiment with a wave tank, a rectangular piece of glass is put diagonally in the path of a series of water waves in the tank.

12.1 What is observed when the waves reach the shallower water over the piece of glass?

12.2 What is this effect called?

12.3 Predict the changes in the
(a) wavelength
(b) the frequency and
(c) the wave speed as the waves move into the shallower water.

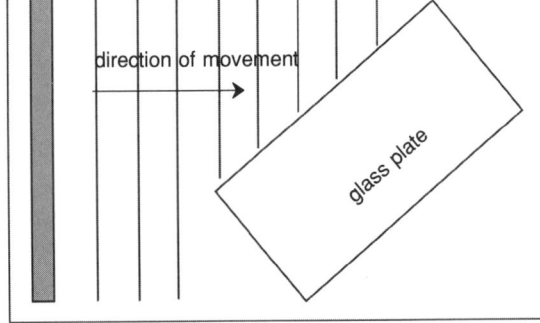

13. How do the speed and wavelength of water waves that approach the beach from the sea change? Explain your answer.

14

14. Two wave series caused by different beads in a wave tank, each with a wavelength of 50 mm, cover 100 mm and 250 mm respectively before they meet. Predict the type of interference that will take place. Explain your answer.

15. Two vibrating beads cause two sets of circular waves in a ripple tank. In the graphical representation of these waves the dotted lines indicate wave troughs and the unbroken lines wave crests.

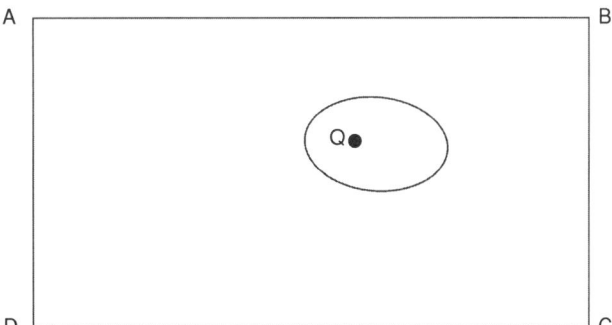

15.1 What is meant by
  (a) constructive interference.
  (b) destructive interference.
15.2 Which of the points A to F represent
  (a) constructive
  (b) destructive interference respectively?

16. A swimming pool ABCD has a shallow and a deep side. If a stone is thrown into the water at point Q, forming a rippling in the water as shown in the sketch.

A ▢ B

Q•

D ▢ C

16.1 Why is the rippling not in a circular form?
16.2 Which side of the swimming pool is the deep side? Explain your answer.

17. Give a name for each of the wave phenomena that are illustrated in the following sketches.

17.1 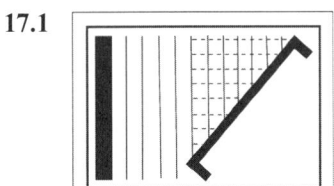 **17.2** 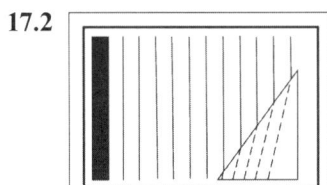 **17.3**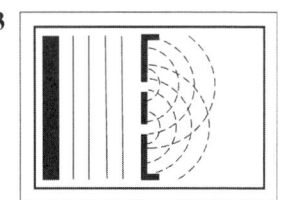

18. Two identical transverse waves A and B move through the same medium. The diagram illustrates the respective positions of the two waves at t = 0 and at t = $\frac{1}{4}$T, where T is the period of each wave.

18.1 What is the speed of the waves if the distance between the vertical lines in the diagram is 2 cm and the period of the waves is 0,1 seconds?

18.2 Name the types of interference that the waves experience at
A. t = 0
B. t = $\frac{1}{4}$T

18.3 Draw the diagrams to illustrate the resulting wave forms during interference at
A. t = 0
B. t = $\frac{1}{4}$T

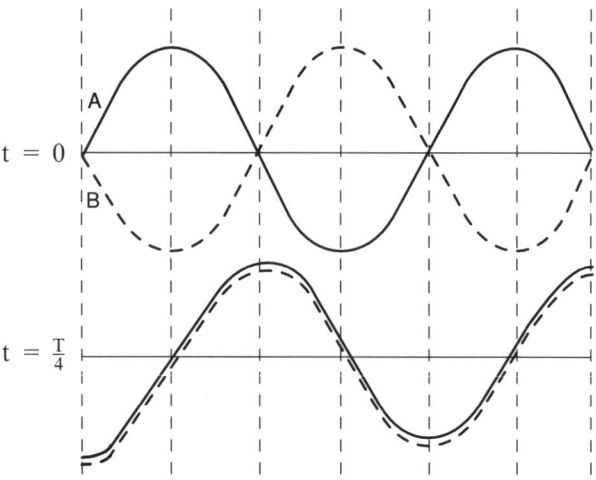

ANSWERS

Section A

1. C	**2.** D	**3.** D	**4.** A	**5.** A	**6.** C	**7.** B
8. C	**9.** C	**10.** B	**11.** A	**12.** C	**13.** A	**14.** A
15. A	**16.** C	**17.** C	**18.** C	**19.** C	**20.** A	**21.** A
22. D	**23.** A	**24.** C	**25.** A			

Section B

1.1 $T = \dfrac{\text{time lapse}}{\text{no. of cycles}} = \dfrac{4}{20} = 0,2 \text{ s}$

1.2 $f = \dfrac{1}{T} = \dfrac{1}{0,2} = 5 \text{ Hz}$

Or: $f = \dfrac{\text{no. of cycles}}{\text{time lapse}} = \dfrac{20}{4} = 5 \text{ Hz}$

2.1 (a) $f = \dfrac{\text{no. of cycles}}{\text{time lapse}} = \dfrac{50}{5} = 10 \text{ Hz}$

(b) $T = \dfrac{1}{f} = \dfrac{1}{10} = 0,1 \text{ s}$

2.2 (a) frequency decreases
(b) period increases

3.1 $f = \dfrac{v}{\lambda} = \dfrac{0,5}{0,025} = 20 \text{ Hz}$

3.2 $T = \dfrac{1}{f} = \dfrac{1}{20} = 0,05 \text{ s}$

3.3 (a) f stays the same $= 20$ Hz

(b) $\lambda = \dfrac{v}{f} = \dfrac{0,4}{20} = 0,02 \text{ m}$

4.1 A and D; B and F

4.2 B; E; F

4.3 distance between E and D

4.4 C

4.5 distance from A to D; or from B to F

5.1 From A to B to C and back to A

5.2 B

5.3 the same as an oscillation (A → B → C → A)

5.4 From B to A or from B to C

5.5 the number of times the movement (A → B → C → A) is repeated per second

5.6 time that it takes to complete one vibration cycle (A → B → C → A)

5.7 maximum displacement from the equilibrium position; i.e. the distance from B to A or from B to C

6.		speed in m.s^{-1}	frequency in Hz	wavelength in m
	Wave 1	32	8	4
	Wave 2	32	16	2
	Wave 3	32	32	1

7.1   period (T) = 0,5 s

∴ time taken for 4 waves = $4 \times 0,5 = 2$ s

7.2 $A = \dfrac{20}{2} = 10$ cm

7.3 $T = 0,5$ s

7.4   (a) $\lambda = 2$ cm $= 0,02$ m

(b) $f = \dfrac{1}{T} = \dfrac{1}{0,5} = 2$ Hz

(c) $v = f\lambda = 2 \times 0,02 = 0,04$ m.s^{-1}

8.1   number of repetitions per second

8.2   transverse waves

8.3   longitudinal waves – direction of propagation of the wave and disturbance of medium are in the same direction (with transverse waves they are perpendicular to each other)

8.4 $\lambda = \dfrac{0,084}{20} = 0,0042$ m

8.5 $v = f\lambda = 10 \times 0,004 = 0,042$ m.s^{-1}

9.1   period (T) $= \dfrac{1}{f} = \dfrac{1}{30}$

∴ time lapse for 3 waves (from A → C) $= 3 \times \dfrac{1}{30} = 0,1$ s

9.2 $\lambda = \dfrac{12}{3} = 4$ m

9.3 $A = \dfrac{4,5}{2} = 2,25$ m

9.4   downwards

9.5 $v = f\lambda = 30 \times 4 = 120$ m.s^{-1}

Or: speed $= \dfrac{\text{distance}}{\text{time}} = \dfrac{12}{0,1} = 120$ m.s^{-1}

10. $f = \dfrac{1}{T} = \dfrac{1}{0,025} = 40$ Hz

$\lambda = \dfrac{v}{f} = \dfrac{12}{40} = 0,3$ m

11.1	speed $= \dfrac{\text{distance}}{\text{time}} = \dfrac{0,4\,\text{m}}{2} = 0,2\ \text{m}\,.\,\text{s}^{-1}$
11.2	$\lambda = \dfrac{0,36}{9} = 0,04\ \text{m}$
11.3	$f = \dfrac{v}{\lambda} = \dfrac{0,2}{0,04} = 5\ \text{Hz}$

12.1	the waves' direction changes (bend towards the glass plate)
12.2	refraction
12.3	(a) decreases (b) stays the same (c) decreases.

13.	the water becomes shallower $\Rightarrow$ the speed decreases and therefore the wavelength also decreases.

14.	constructive interference; the difference in the distance that they cover is $250 - 100 = 150$ mm, which is equal to precisely 3 wavelengths. The two waves are in phase and two crests or two troughs meet.

15.1	(a) two waves in phase that meet and reinforce each other. (b) two waves out of phase that meet and cancel each other out.
15.2	(a) A, B, E (b) C, D, F

16.1	the swimming pool's depth is not the same all over.
16.2	BC – the waves move faster to this side $\Rightarrow$ therefore the water must be deeper.

17.1	reflection
17.2	refraction
17.3	interference

18.1	$f = \dfrac{1}{T} = \dfrac{1}{0,1} = 10\ \text{Hz}$ $v = f\lambda = (10)(4 \times 0,02) = 0,8\ \text{m}\,.\,\text{s}^{-1}$
18.2	A: destructive interference B: constructive interference
18.3	A: $t = 0$ B: $t = \tfrac{1}{4}T$

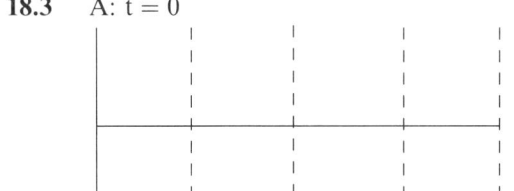
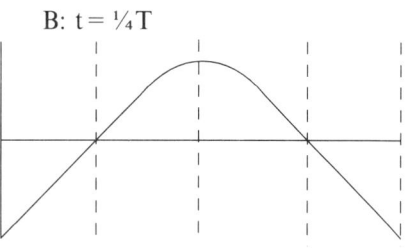

2 *Refraction of Light*

1. Refraction

1.1 What is refraction?

As in the previous chapter, light, because it moves as a wave, can be **bent** or **refracted**. This phenomenon is illustrated when a straight stick is immersed obliquely in a bowl of water and it looks as if the stick is sligthly bent.

When light moves from one optic medium to another medium with a different optical density, the speed of the light wave changes. If a light ray strikes the dividing line between two media at a slant (obliquely), the change in the speed of the wave will cause a change in direction of the ray.

When light moves from a medium with an **optically lower** to a medium with an **optically higher** density (from air to glass), and strikes the dividing line obliquely, the wave moves **slower**, and the ray is bent **closer to the normal**. The angle of incidence is **greater** than the angle of refraction ($\angle$ i > $\angle$ r).

When light moves obliquely from a medium with an **optically higher to an optically lower density** (from glass to air), the opposite is true and the wave moves **faster** so that the ray is bent **away from the normal**. The angle of incidence is **smaller** than the angle of refraction ($\angle$ i < $\angle$ r).

When a light ray moves **perpendicularly** from one medium to another, **no refraction** takes place.

1.2 Refraction through a rectangular glass block

When a light ray is shone through a rectangular glass block, the refraction that takes place can be presented by the following sketch.

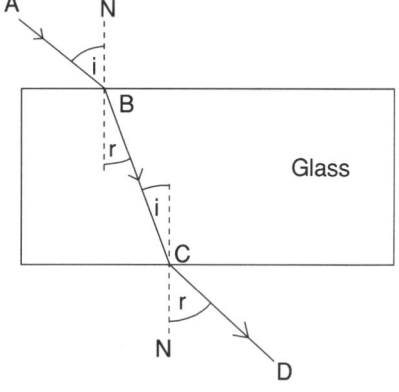

Terms

- AB is the **incident ray.**
- BC is the **refracted ray.**
- CD is the **emerging ray.**
- The angle between the incident ray and the normal is the **angle of incidence ($\angle$ i).**
- The angle between the refracted ray and the normal is the **angle of refraction ($\angle$ r).**

As glass has a higher optical density than air, the light ray moves slower in it and is therefore bent nearer to the normal at the first refractive surface ($\angle$ i > $\angle$ r). At the second refractive surface, where the light ray moves from glass to air, the opposite occurs so that the ray is bent away from the normal ($\angle$ i < $\angle$ r). The emerging ray is then parallel to the incident ray.

1.3 Refraction through a triangular prism

When a light ray is shone obliquely through a triangular glass prism, the light ray is bent nearer to the normal at the first refractive surface (from air to glass) and away from the normal at the second refractive surface (from glass to air). The light ray is bent towards the base of the prism.

It can be represented as follows on the accompanying sketch:

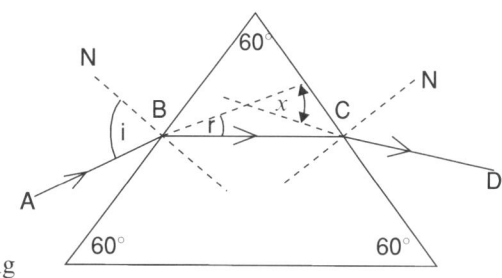

Terms

- Similar to the refraction at a rectangular glass block, AB is the **incident ray**, BC the **refracted ray**, CD the **emerging ray**, $\angle$ **i** the **angle of incidence** and $\angle$ **r** the **angle of refraction**.
- $\angle$ **x** is the **angle of deviation** (the angle of deflection between the incident and emerging rays)

When the refracted ray is parallel to the base of the prism, it is found that the emerging ray deflects the least from the incident ray and the angle of deviation is a minimum. The prism is then in the position of minimum deviation.

1.4 True and apparent depth

When one looks at an object in a bowl of water, it looks as if the water is shallower than what it actually is. This phenomenon is called apparent depth and is due to light refraction.

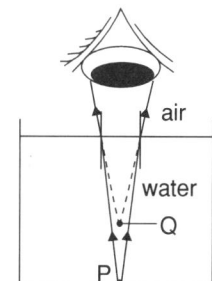

The diagram shows a light ray from object P at the bottom of a bowl of water, that is refracted at the surface of the water. Seeing that the light ray moves from the optically dense water to optically less dense air, it is bent away from the normal.

A person's eye can only see in straight lines and can therefore not observe the change in direction of the ray. A virtual or apparent image is observed by the eye at point Q, which is formed above the object.

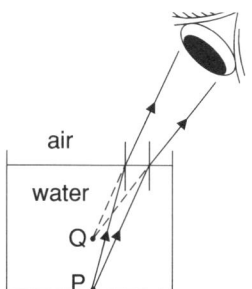

2. Total internal reflection

2.1 What is total internal reflection?

When light moves obliquely from an optically denser medium to an optically less dense medium, it is bent away from the normal with the result that the angle of refraction is greater than the angle of incidence. However, if the angle of incidence is made systematically greater, the stadium will be reached where the refracted ray skims the dividing surface (at (c) in the sketch).

If the angle of incidence is increased further, the angle of refraction becomes greater than 90° and no more refraction takes place. The light ray is reflected back into the optically denser medium (at (d) in the sketch). This phenomenon is called **total internal reflection**.

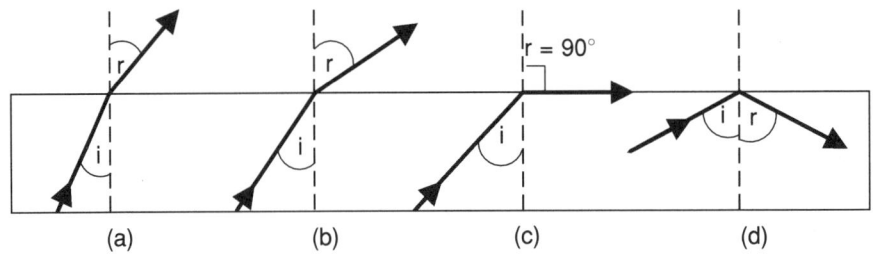

(a)	(b)	(c)	(d)

The angle of incidence for which the light moves **along the dividing surface** (at (c) in the sketch), is called the **critical angle**.

The critical angle is therefore the magnitude of the angle of incidence which causes an angle of refraction of 90°.

Conditions for total internal reflection:
- Light must move from an optically dense medium to an optically less dense medium;
- The angle of incidence must be greater than the critical angle for the medium involved.

Example:

The critical angle for water is 49°. If a light ray moves from water to air with an angle of incidence of 45°, refraction will take place. However, if the angle of incidence is increased to 50°, all the light will be reflected back into the water, with the angle of incidence equal to the angle of reflection.

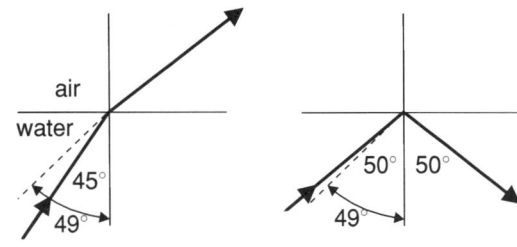

2.2 Applications of total internal reflection

- **Periscopes and Binoculars:**

  Prisms which can be used as mirrors in objects such as periscopes and binoculars have many more advantages than normal mirrors. Prisms reflect almost all light and further have no reflecting surface that can be eroded.

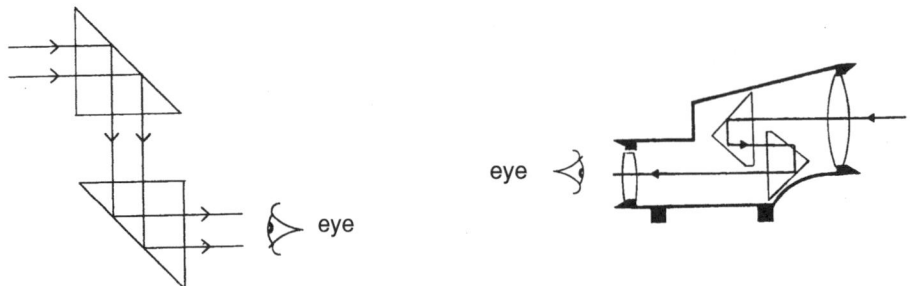

45° prisms can be used as mirrors to change the direction of light rays by reflecting them in different ways. The different possibilities are shown in the diagrams below.

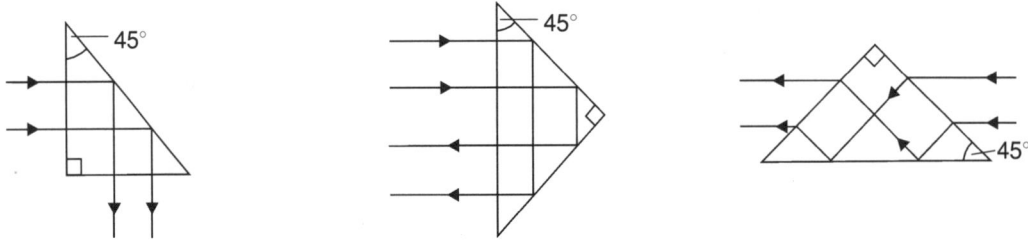

- **Multiple Internal Reflection**

  A bunch of thin glass fibres can be used as a "light pipe". The light rays are caught up within the different fibres by means of total internal reflection. The light rays are reflected repeatedly between the walls of each fibre and are as such conducted through the fibres.

3. The colour spectrum

3.1 Dispersion

When white light is shone through a prism, it is broken up into a rainbow of seven colours, namely red, orange, yellow, green, blue, indigo and violet. This rainbow is called the visible spectrum. When this rainbow is once again joined, white light is obtained.

The **separation of white light into its constituent colours** is called **dispersion**. **Violet light** is **refracted** the **most** and **red light** the **least**. Violet light moves slowest through the prism, whereas red light moves the fastest.

3.2 The colour of objects

An object shows a specific colour when white light, which is a mixture of different colours, falls on it. When light falls on an object, it can be **reflected**, **absorbed** or **transmitted**.

By suitable mixing of the three light colours, **red, green and blue**, any colour can be obtained. The three colours are known as **primary colours.**

Secondary colours are obtained by mixing two or three primary colours.

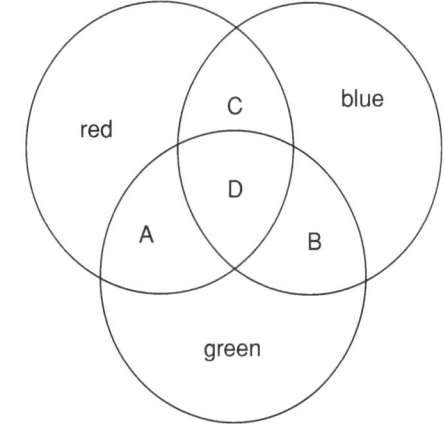

- **Yellow (A)** is obtained by mixing red and green;
- **Cyan (B)** by mixing blue and green;
- **Magenta (C)** by mixing red and blue;
- **White light (D)** by mixing red, blue and green.

3.3 Opaque objects

An opaque object has colour when it reflects some colours of white light and absorbs others.

When an **object absorbs all colours, it appears black.** Black can be regarded as the absence of light.

An object appears white if white light is shone onto it and it reflects **all colours** (red, blue, green). A white object will therefore appear red when red light is shone onto it because the red light is reflected; appears blue in blue light and appears green in green light.

A blue object reflects only blue light and absorbs all other colours. It will therefore only appear blue in white or blue light and black in all other colour light. This phenomenon is known as **selective reflection.**

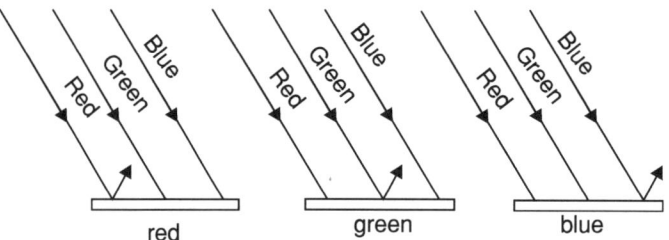

red   green   blue

A yellow object will appear yellow in white or yellow light, but because yellow is a secondary colour which consists of a mixture of green and red light, it will appear green in green light and red in red light. In blue light it will appear black because the blue is absorbed.

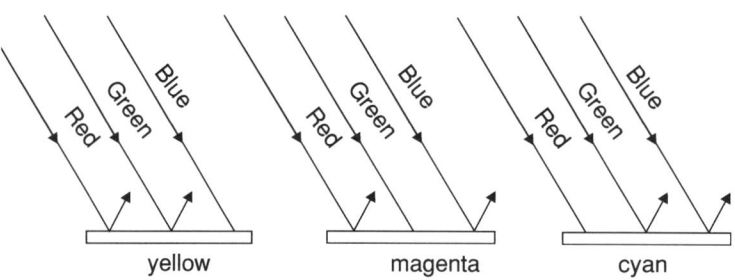

yellow   magenta   cyan

3.4 Transparent objects

The **colour of transparent objects** depends on the **colour light that they transmit.** They transmit their own particular colour and absorb the other colours. This phenomenon is known as **selective absorption.** A transparent coloured substance therefore acts as a **colour filter.**

Blue glass acts as a filter and allows only blue light through while the other colours are absorbed. Similarly a red filter transmits only red light and a green filter only green light. When two primary colour-filters are placed on top of each other, the light that is transmitted through the one filter is absorbed by the other, so that no light is transmitted and appears to be black.

A yellow filter transmits yellow light, but also red and green light (yellow is a mixture of red and green).

QUESTIONS

Section A

Different possibilities are proposed as answers for the following questions. Choose the correct answer.

1. A light ray strikes the straight side of a semi-circular glass block so that it strikes the middle of the straight side obliquely. Which one of the following diagrams illustrates the path of the ray through the glass block?

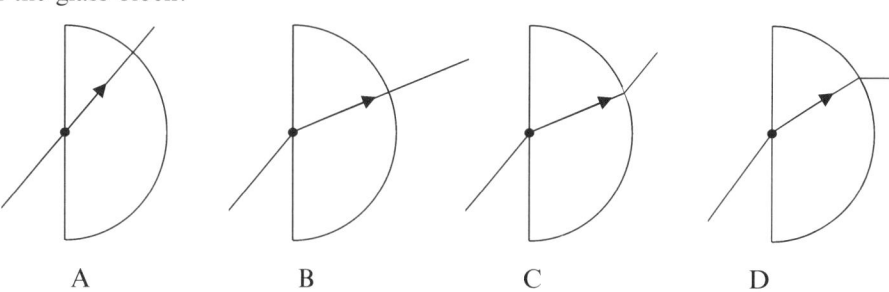

| A | B | C | D |

2. When a light ray enters a glass block obliquely, its direction changes. This phenomenon is known as . . .
   A   diffraction.   C   dispersion.
   B   refraction.   D   aspersion.

3. If a stick is immersed in water, it appears as if the point bends . . .
   A   down, because light undergoes refraction.
   B   down, because light is reflected.
   C   up, because light undergoes refraction.
   D   up, because light is reflected.

4. A diver near the surface of the water, sees a fish in deeper, much colder sea water as shown in the sketch.
   If he wants to hit the fish,
   he must aim at a point . . .

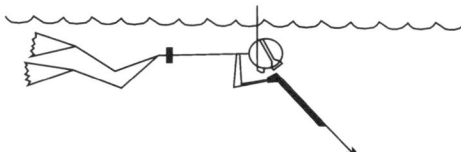

   A   above the fish.
   B   below the fish.
   C   behind the fish.
   D   on the fish.

5. A light ray that strikes the dividing surface between air and water perpendicularly, . . .
   A   is bent towards the normal.   C   is reflected.
   B   is bent away from the normal.   D   doesn't change direction.

6. When a light ray from the air strikes the dividing surface between air and water obliquely, the ray is . . .
   A   refracted to skim the surface of the water.
   B   refracted away from the normal.
   C   refracted towards the normal.
   D   reflected according to the rules of total internal reflection.

7. Which of the following sketches is correct for light which moves from air to glass?

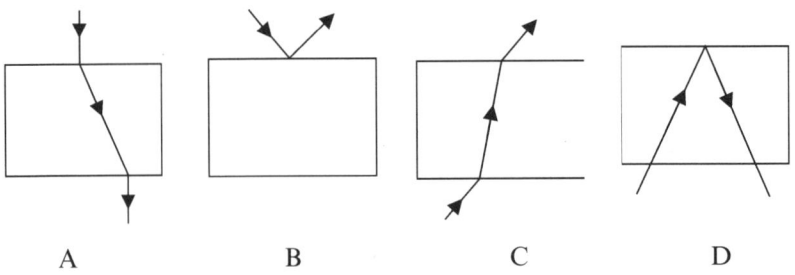

A   B   C   D

8. At which point in the sketch will the coin be observed?

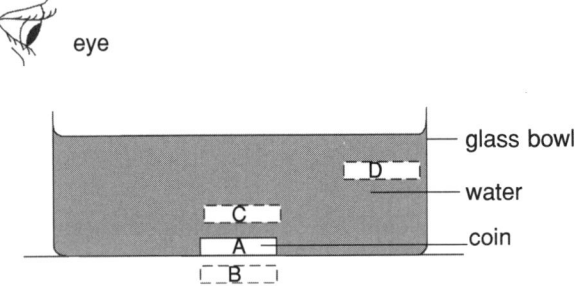

9. Which one of the following light ray diagrams indicates why a bowl of water seems more shallow than it really is?

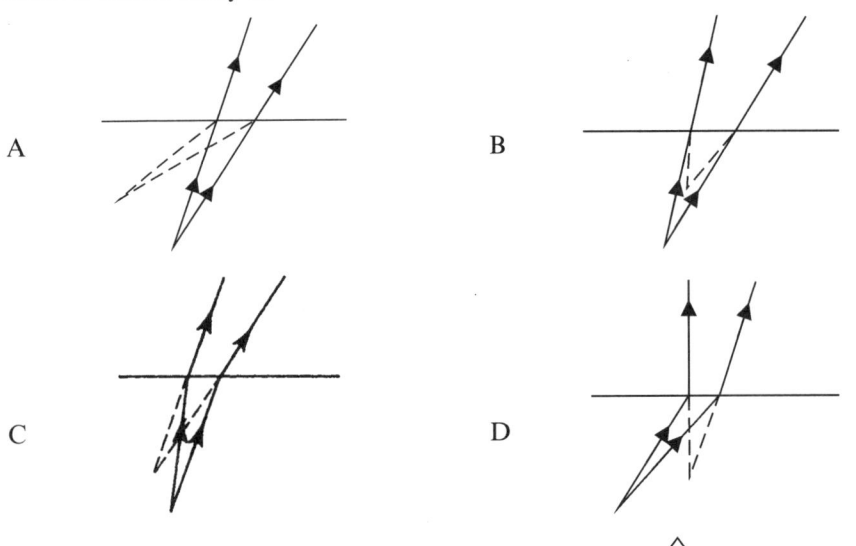

10. Which angle represents the angle of deviation in the following diagram?

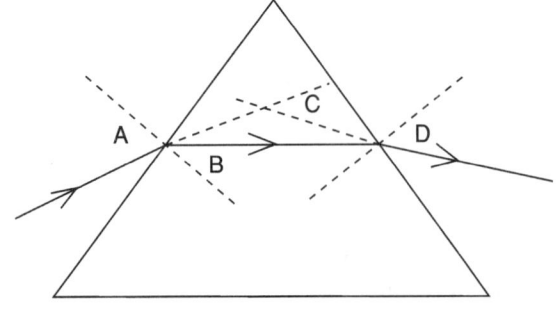

11. Which statement is **false**?

When a light ray is refracted through a 60° prism and minimum deviation occurs, the . . .

A   angle of incidence at the first refractive plane is equal to the angle of refraction at the second refractive planes.
B   refracted ray is parallel to the base of the prism.
C   deviation of the emerging ray from the incident ray is a maximum.
D   magnitude of the angle of refraction is a minimum.

12. The critical angle of perspex is 42°. Which one of the following sketches correctly shows the path of a light ray through perspex?

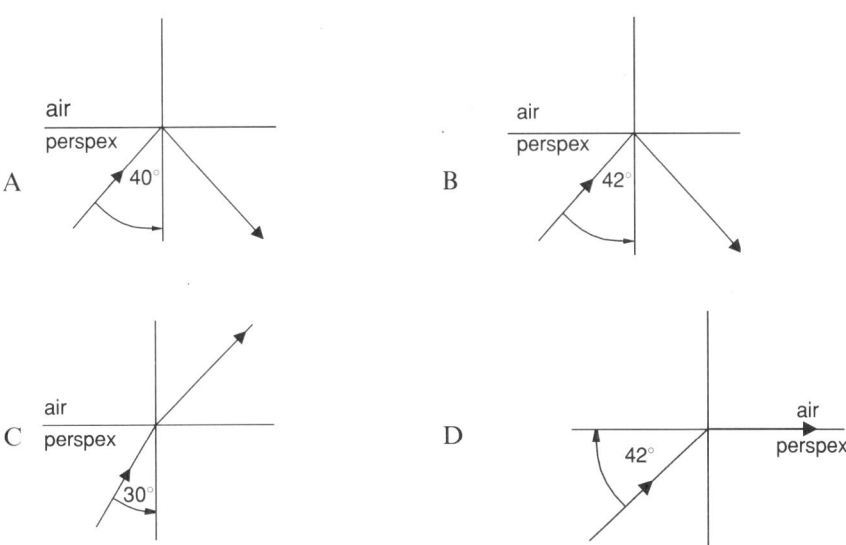

13. The critical angle of water with regard to air is 49°. Total internal reflection will only take place if the angle between the ray that moves through the water and the dividing surface is equal to . . .

A   30°   C   60°
B   49°   D   90°

14. Total internal reflection will occur when light moves from . . .

A   water to air and the angle of incidence is larger than the critical angle.
B   water to air and the angle of incidence is smaller than the critical angle.
C   air to water and the angle of incidence is greater than the critical angle.
D   air to water and the angle to incidence is smaller than the critical angle.

15. Which one of the following can be the angle of refraction when light enters glass from air with an angle of incidence of 30°?

A   90°   C   45°
B   30°   D   25°

16. The critical angle of light refraction is . . .

A   the angle of incidence which gives an angle of reflection of 90°.
B   the angle of incidence which gives an angle of refraction of 90°.
C   the angle of refraction that is formed when light strikes the dividing surface of two media at 90°.
D   the angle of refraction that is formed when light strikes parallel to the dividing surface between two media.

17. When total internal reflection occurs, the magnitude of the angle of reflection is . . .

 A   smaller than the angle of incidence.   C   the same as the angle of incidence.

 B   greater than the angle of incidence.   D   the same as the critical angle.

18. Which one of the following sketches indicates the path of a light ray at the top of a periscope?

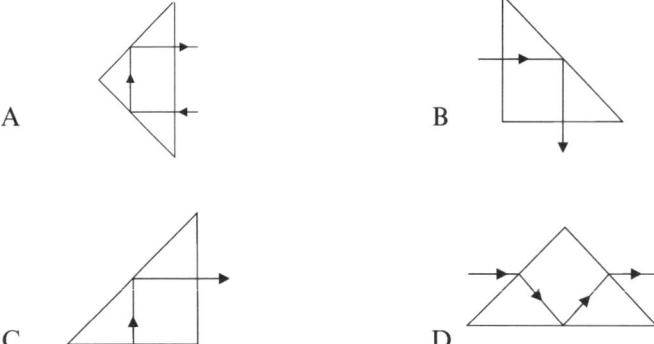

A   B

C   D

19. In which sketch is ∠ x the critical angle?

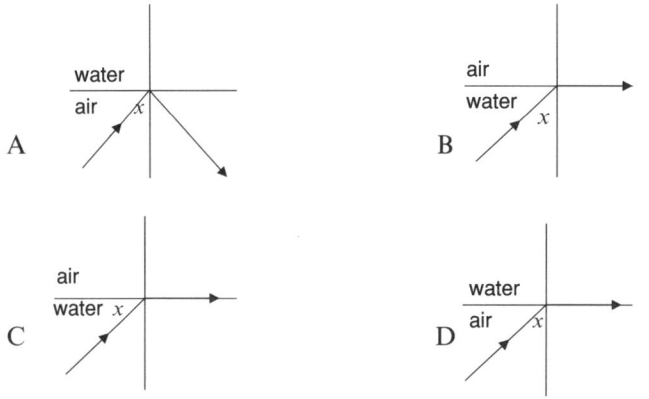

A   B

C   D

20. The incident ray and emerging ray don't have the same direction if a red light ray moves through a 60° prism. The angle between the incident ray and the emerging ray is called the . . .

 A   angle of incidence.   C   angle of reflection.

 B   angle of deviation.   D   angle of refraction.

21. Which one of the following phenomena is **not** an illustration of light refraction?

 A   Red light that moves through a prism   C   True and apparent depth

 B   Mirages   D   Green light that is absorbed by red glass

22. If white light is observed through green glass, the green glass will . . .

 A   reflect green light.   C   reflect red and blue light.

 B   absorb green light.   D   absorb red and blue light.

23. A red book that is illuminated by green light, will . . .

 A   appear red.   C   appear black.

 B   appear green.   D   appear yellow.

24. An object is black when it . . .

 A   absorbs all colours of light.   C   transmits all colours of light.

 B   reflects all colours of light.   D   reflects black light.

25. When a green light ray is shone onto a sketch made with a turquoise pen, the sketch appears . . .
A green.
B turquoise.
C black.
D blue.

26. Three projectors, with red, blue and green colour filters respectively, are directed at a screen.

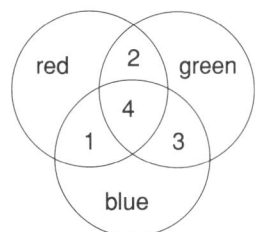

The areas 1, 2, 3 and 4 where the bundles overlap, are respectively . . .

	1	2	3	4
A	magenta	yellow	turquoise	white
B	yellow	magenta	turquoise	grey
C	magenta	turquoise	yellow	white
D	turquoise	yellow	magenta	white

Section B

1. Draw a labelled ray diagram to illustrate the path of a light ray through
   1.1 a rectangular glass block.
   1.2 a triangular glass prism where minimum deviation takes place.

2. Five parallel light rays strike a glass sphere as shown in the sketch. Redraw the sketch and complete to illustrate the path of the light rays through the sphere.

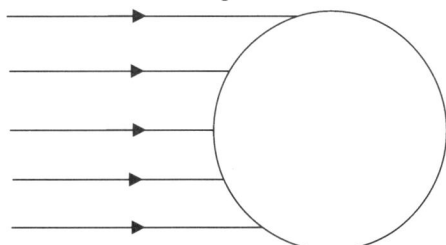

3. Copy and complete the following diagrams to illustrate the path of the light ray:

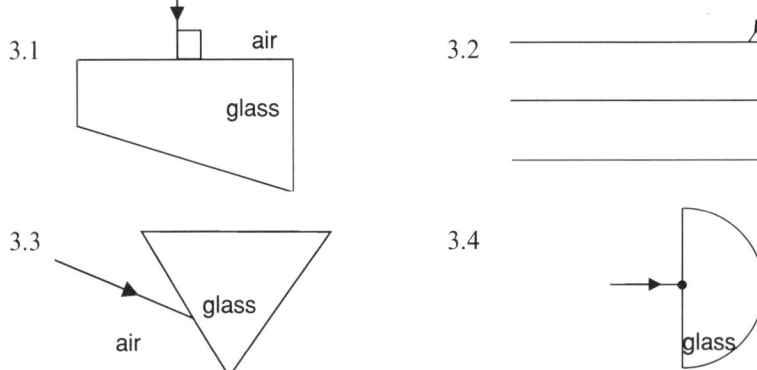

3.1

3.2

3.3

3.4

4. A stone lies at the bottom of a dam. A boy stands on the dam wall and looks at the stone. Illustrate, with the help of a ray diagram, where he will observe the stone.

5. Figure A illustrates the image that a fish in a dam of water will see if it looks up straight at a person outside the water. Figure B illustrates a ray diagram of the light rays to the fish's eye.

Figure A

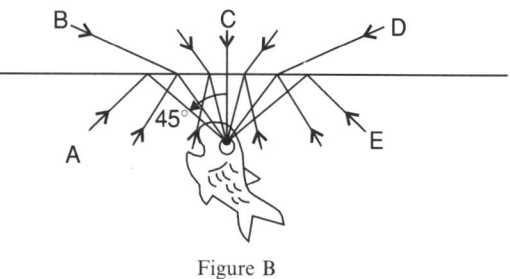

Figure B

Explain the shape of the image that would form in a fish's eye by referring to what happens to rays A to E.

6. The figure illustrates how a person who is visiting an aquarium, sees a seal in a tank of water.

Give an explanation for what occurs here.

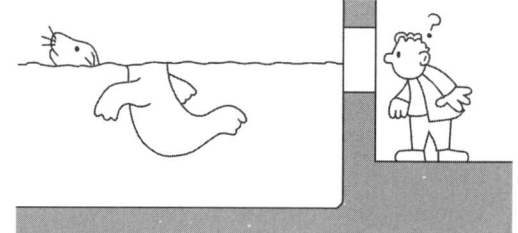

7. Sea birds instinctively know to dive for a fish from directly above it. Why is this so?

8. In the following two sketches the glass block is exactly as big as the volume of water in the glass bowl. In both cases, the observer sees an apparent image of the coin when he looks from directly above it.

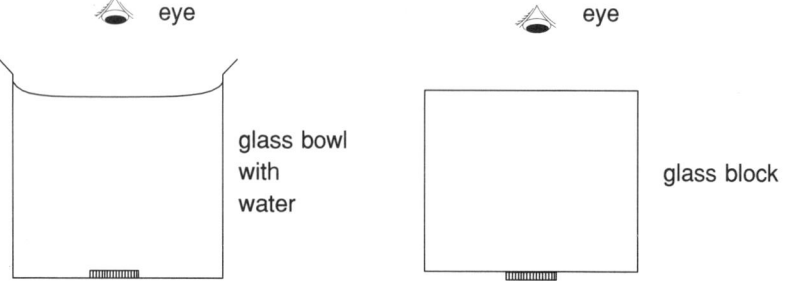

glass bowl with water

glass block

8.1 In which case will the distance between the image that is observed and the coin be the greatest?

8.2 Explain your answer by referring to optical density and the amount of refraction.

9. A prism is needed to reflect a light bundle through an angle of 90°. One of the following prisms can be used:

I. a hollow 45° prism that is filled with water (the critical angle of water is 49°).
II a hollow 45° prism that is filled with carbon disulphide (the critical angle of carbon disulphide is 38°).

9.1 Which choice will be the best?

9.2 Explain your answer.

10. In order to glitter, a diamond is cut in such a way so that the light that strikes it is "caught" within the diamond. Explain how this is possible.

11. The sketch illustrates how two prisms are incorrectly used to make a simple periscope.

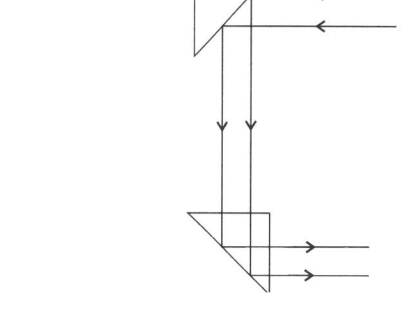

11.1 Redraw the diagram correctly and show the angles of incidence, refraction and reflection.
11.2 Why are prisms used instead of mirrors when a good quality periscope is required?

12. The path of a single ray of white light through a 60° prism is examined.

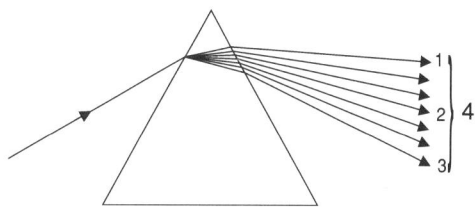

12.1 What do the following numbers represent?
A. 1
B. 3
C. 4

12.2 What phenomenon does the above sketch illustrate?

12.3 Numbers 1, 2 and 3 are known as the . . . of white light.

12.4 Which colour light moves with the greatest speed through the prism? Give an explanation for your answer.
12.5 What is observed if a red filter is placed in the path of the light ray in front of the prism? Give an explanation for your observations.
12.6 If the red filter is replaced with a blue filter, how will your observation differ?
12.7 A drawing in green ink on white paper is now illuminated with light from number 1. What colour will the paper and the sketch be respectively? Explain.

13. A woman tries on a yellow dress with a red belt in the following colour light:
13.1 Red light
13.2 Blue light
13.3 Green light

What will the colours of the dress and the belt be respectively in each case?

14. A red light ray passes through a 60° prism and is then shone onto a piece of yellow cardboard.

    14.1 What is observed regarding the
        A. prism and
        B. the cardboard?

    14.2 What do we call the phenomenon that is observed regarding the
        A. prism and
        B. the cardboard?

    14.3 How do the observations change in the case of the
        A. prism and
        B. cardboard if the red light is replaced by blue light?
        Explain your answers.

15. What colour will a TV-set's screen show when the following colours are mixed?

    15.1 Red and green;

    15.2 Red, green and blue;

    15.3 Red and blue.

ANSWERS

Section A

1. B	**2.** B	**3.** C	**4.** B	**5.** D	**6.** C	**7.** C
8. C	**9.** C	**10.** C	**11.** C	**12.** C	**13.** A	**14.** A
15. D	**16.** B	**17.** C	**18.** B	**19.** B	**20.** B	**21.** D
22. D	**23.** C	**24.** A	**25.** A	**26.** A		

Section B

1.1

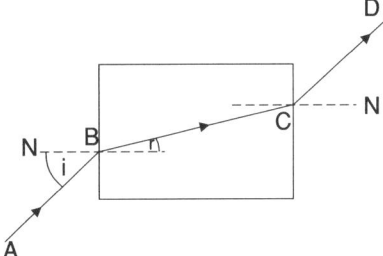

N – normal
AB – incident ray
BC – refracted ray
CD – emerging ray
∠ i – angle of incidence
∠ r – angle of refraction

1.2

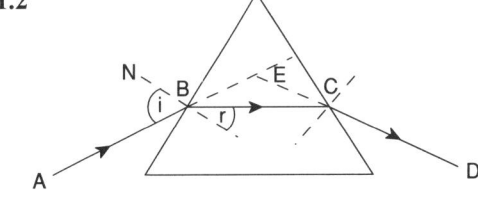

N – normal
AB – incident ray
BC – refracted ray
CD – emerging ray
∠ i – angle of incidence
∠ r – angle of refraction
∠ E – angle of deviation

2.

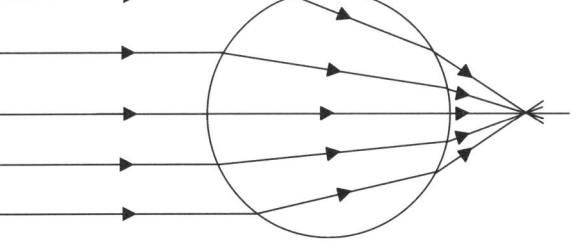

3.1

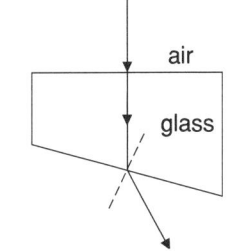

3.2

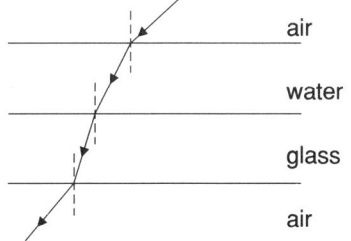

33

3.3.

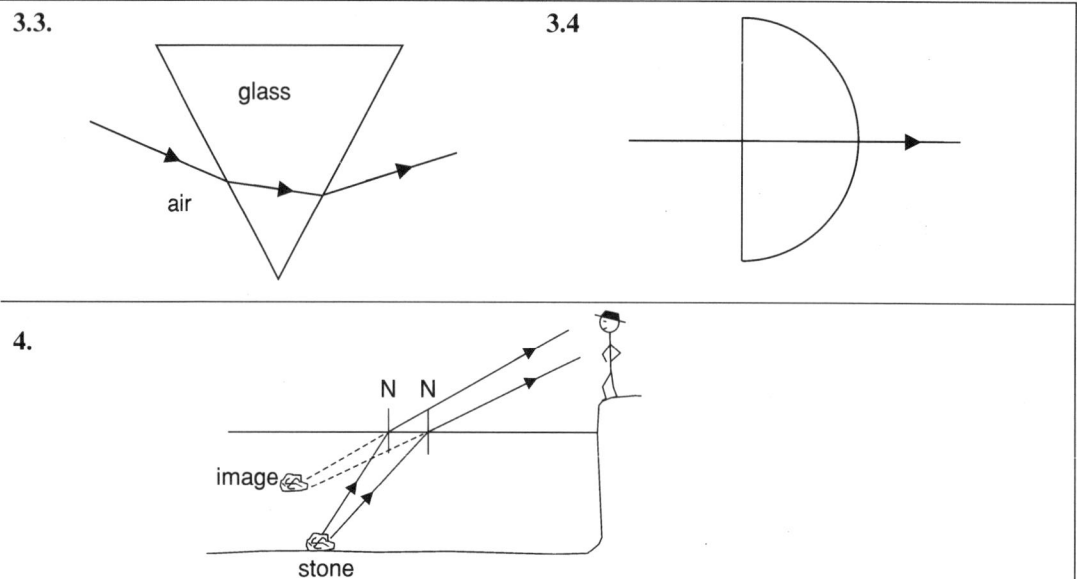

3.4

4.

stone

5.

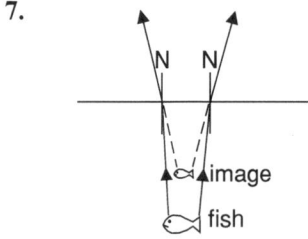

The fish looks up through a cone that forms an angle of 48,5° to each side (the critical angle of water). With larger angles, rays from under the water that undergo total internal reflection, strike the fish's eye. However the fish still sees the total picture outside the water with the help of rays B to D. The edges of the image are however distorted, as rays B and D are refracted. The middle of the image is seen clearly, as the rays at C are not refracted.

6. Due to apparent depth, the part of the seal that is in the water appears closer to the person than the part that is in the air, where no air refraction takes place.

7.

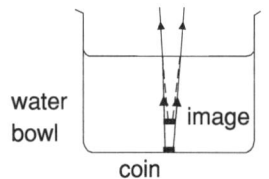

Although the image is shallower than the fish, the fish is still directly below it. If the bird looks at it from the side, he will dive at the image above the fish and therefore misses the fish.

8.1 at the water bowl

8.2 The water is less dense than the glass block. There is therefore less refraction between the water and the air than between the glass block and the air. The apparent image that forms in the water is "deeper" than in the glass block. The image in the glass block is therefore the closest to the observer's eye.

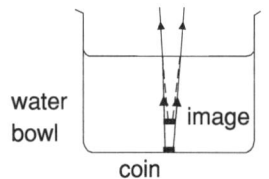

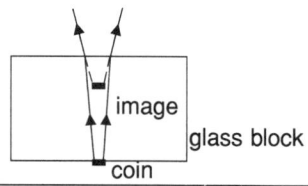

9.1 the prisma filled with carbon disulphide

9.2

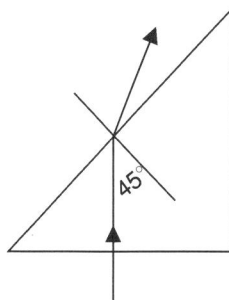

For a prism to rotate a light ray through 90°, it must be set up as in the sketch; the angle of incidence at the reflecting surface must therefore be 45°. Water's critical angle is 49°; therefore the light that strikes the dividing surface between air and water at an angle of 45°, will be refracted and not reflected. Carbon disulphide's critical angle is 38°, which is smaller than 45°, therefore total internal reflection will take place here.

10. The surfaces are cut in such a way so that light that penetrates one of the surfaces, will strike the other surfaces at angles greater than the critical angle for diamond. Total internal reflection takes place continually.

11.1

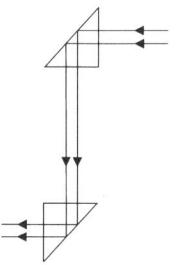

11.2 It reflects all light, whereas mirrors only reflect about 70% of incoming light.
– Prisms do not have reflecting surfaces that can erode.

12.1 A: red light
B: violet light
C: colour spectrum

12.2 dispersion

12.3 primary colours

12.4 red; it is refracted the least

12.5 only red light is transmitted. Due to selective absorption, the filter absorbs the other colours of white light and only transmits red light.

12.6 only blue light is transmitted

12.7 the sketch appears black; the green sketch absorbs the red light which falls on it (selective reflection); the paper appears red; white reflects all colour light that falls on it because white is a combination of red, blue and green.

13.1 dress: red
belt: red

13.2 dress: black
belt: black

13.3 dress: green
belt: black

14.1 A: light ray is bent
B: cardboard appears red

14.2 A: refraction
B: selective reflection

14.3	A: blue light is bent more; it undergoes more refraction because it travels slower than red light through the prism.
	B: cardboard appears black; yellow is a combination of red and green light; it therefore absorbs the blue light that falls on it.
15.1	yellow
15.2	white
15.3	magenta

3 *Lenses*

1. Terminology

- A **lens** is any transparent substance that is bordered by at least one curved surface. It can be considered as a composition of parts of prisms through which light is refracted in keeping with the laws of refraction.

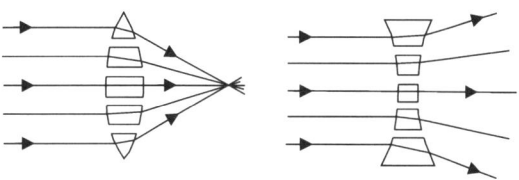

- A **convex lens** makes light rays converge and is curved to the outside, so that it is thicker in the middle than at the edges.

- A **concave lens** makes light rays diverge and is curved to the inside, so that it is thinner in the middle than at the edges.

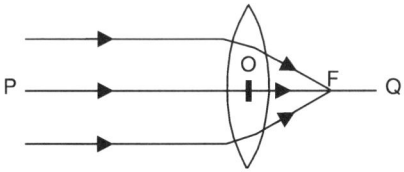
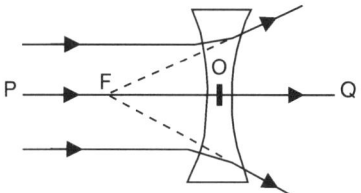

- The **centres of curvature** of a lens are the centres of the two circles of which the surfaces of the lens form a part.

- The **principal axis** of a lens (line PQ on the sketches) is a line that joins the centres of curvature of the lens.

- The **optical centre** (point O on the sketches) is the point in the lens through which light moves without changing direction.

- The **focal point** (point F on the sketches) of a lens, is the point where rays that strike a convex lens parallel to its axis, converge, and the point from which rays that strike a concave lens parallel to its axis, apparently diverge.

- The **focal length** (the distance OF on the sketches) is the distance between the optical centre and the focal point of the lens.

2. Image forming through lenses

2.1 Ray-diagrams

To examine the images that are formed by lenses, we use **ray-diagrams**.
To draw a **ray-diagram for a convex lens**, we make use of the following three rays:
- a Ray parallel to the principal axis that is refracted through the focal point;
- a Ray that moves through the optical centre of the lens without being refracted;
- a Ray passing through the focal point in front of the lens, that emerges, after being refracted, parallel to the principal axis.

The two rays that are used to draw a **ray-diagram for a concave lens**, are
- a ray parallel to the principal axis that appears to diverge from the focal point;
- a ray that moves through the optical centre of the lens without being refracted.

In ray-diagrams a vertical line is drawn through the optical centre of the lens and light rays are drawn in such a way that it seems as if refraction occurs there.

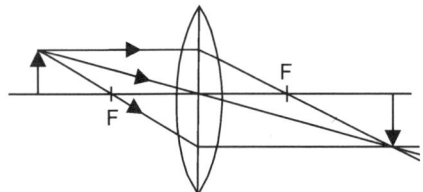
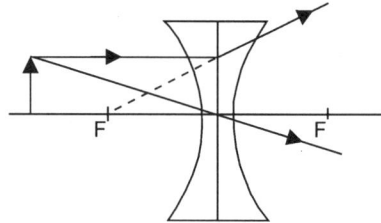

2.2 Image forming

The extent to which a lens refracts light, depends on its focal length. Lenses with short focal lengths bend the light rays more so that larger images are formed than with lenses with longer focal lengths. The size of the image that is formed by a lens, is also influenced by the distance of the object from the lens. Images can be real or virtual.

A real image
- is formed when the rays actually pass through the image;
- can be projected onto a screen;
- is always inverted in relation to the object.

A virtual image
- is formed when the rays do not actually pass through the image;
- cannot be projected onto a screen;
- is always upright in relation to the object.

2.3 Image forming with convex lenses

Different observations are made when an object is placed at different distances from a convex lens. When the object is placed far from the lens, a real, reduced and inverted image is formed. As the object is placed nearer to the lens, the image becomes larger. When the object is placed at precisely two times the focal length (2F) from the lens, the image that is formed is just as big as the object. When the object is placed between the lens and the focal point, a virtual, upright and enlarged image is formed.

A summary of the different possible situations is supplied in the following diagram:

Ray-diagram	Characteristics of image	Practical uses
Object very far from lens 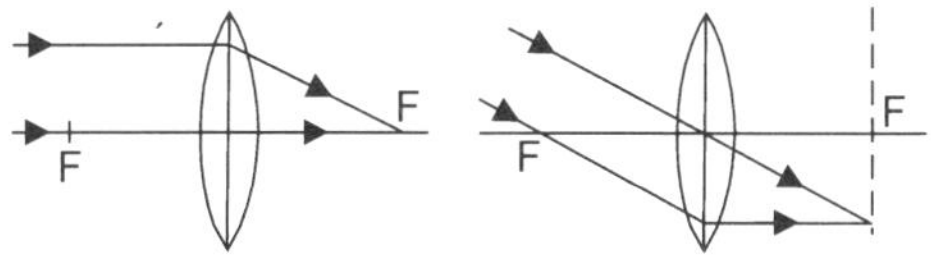	real; a light point on the focal point or the focal plane	photography of stars or distant landscapes
Object further than 2F from the lens 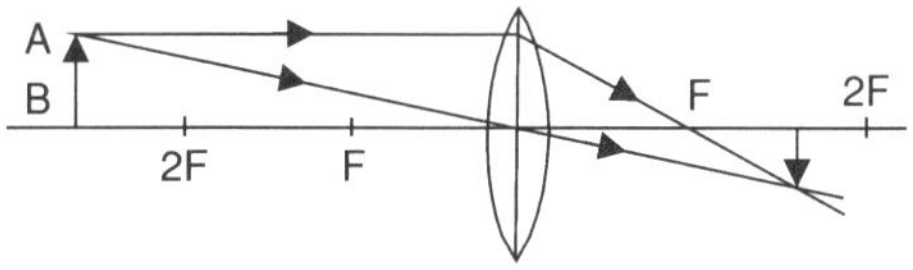	real; reduced; inverted	camera lens; eye; objective in binoculars and telescopes
Object on 2F of the lens 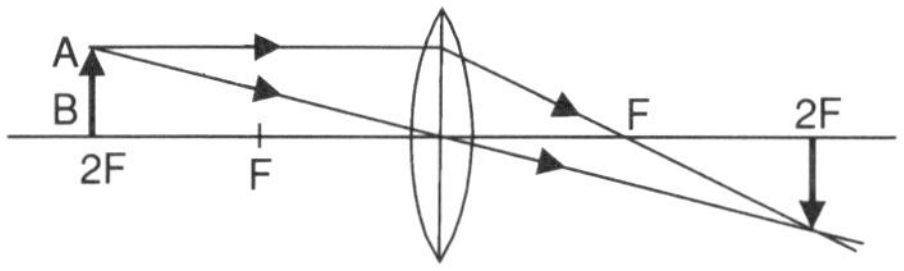	real; same size as object; inverted	copier camera; to obtain an upright image in a sliding telescope
Object between 2F and F from the lens 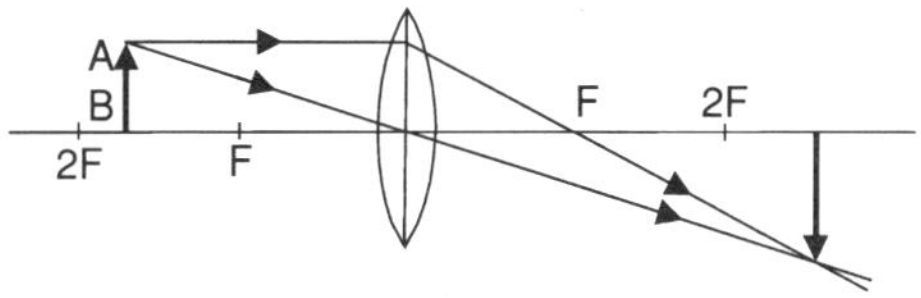	real; enlarged; inverted	film projectors; photographic zoom lens; microscope objectives
Object on F of the lens 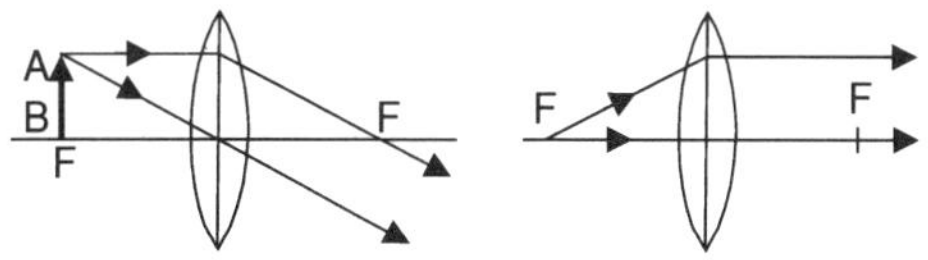	no image	spotlights; searchlights; collimators
Object between F and the lens 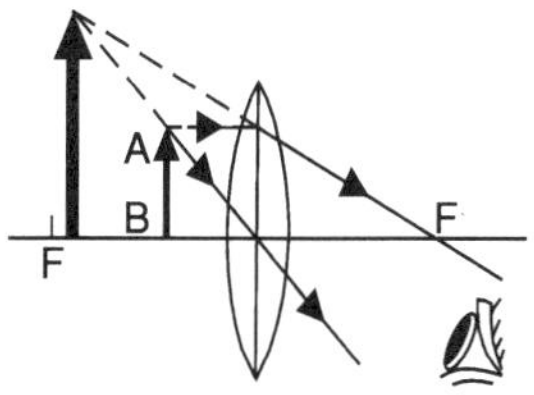	virtual; enlarged; upright	magnifying lens; eyepiece for telescopes and microscopes

2.4 Image forming with concave lenses

With concave lenses, the image that is formed is **always** upright, reduced, virtual and on the same side of the lens as the object.

Concave lenses are generally used to increase the focal length of convex lenses (e.g. glasses' lenses for near-sighted people).

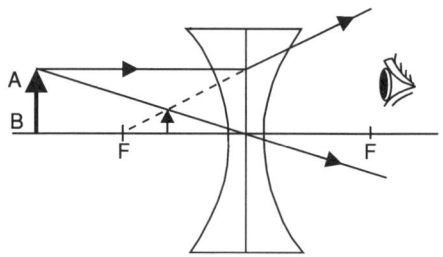

2.5 Graphic constructions

If the object distance and the focal length are known, the position and size of the image that is formed can be determined with an accurate construction. Consider the following examples:

Example 1:

An object with a size of 20 mm is placed 65 mm from a convex lens. The lens's focal length is 25 mm. Determine the size of the image that is formed.

Solution:

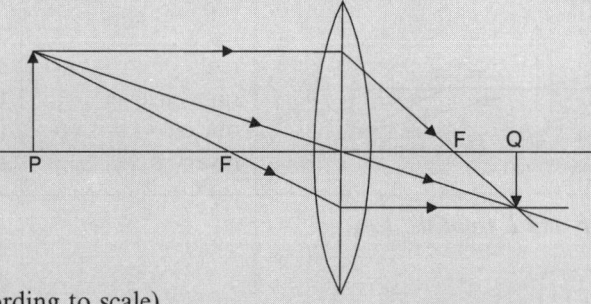

(Construction not according to scale)
Image size = 12 mm

Example 2:

An object, 20 mm high, is placed 80 mm from a concave lens with a focal length of 50 mm. Determine the size of the image that is formed, by means of an accurate construction.

Solution:

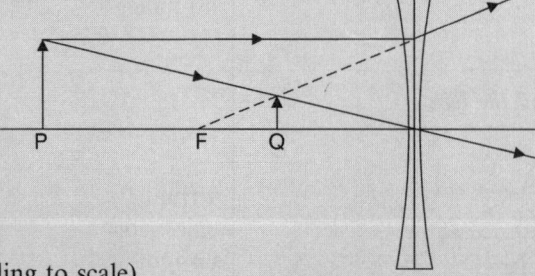

(Construction not according to scale)
Image size = 8 mm

3. Magnification

The magnification of an image (i.e. how many times the image is larger than the object) can be calculated with the following equation:

$$\text{magnification} = \frac{\text{image size}}{\text{object size}} = \frac{\text{image distance}}{\text{object distance}}$$

Seeing that the magnification is expressed as an equation, it has no unit. If the magnification is 1, the image and the object are the same size. If the magnification is greater than 1, the image is larger than the object, and if the magnification is less than 1, the image is smaller than the object.

Example 1:

An object with a size of 15 mm is seen through a magnifying glass. If an image with a size of 25 mm is formed 28 mm from the lens, calculate
a. the magnification and
b. the object distance.

Solution:

a. $\qquad$ Magnification $= \dfrac{\text{image size}}{\text{object size}} = \dfrac{25}{15} = 1,67$

b. $\qquad$ Magnification $= \dfrac{\text{image distance}}{\text{object distance}}$

$$\therefore 1,67 = \frac{28}{\text{object distance}}$$

$$\therefore \text{object distance} = \frac{28}{1,67} = 16,8 \text{ mm}$$

Example 2:

An object, 20 mm high, is placed 45 mm from a convex lens with a focal length of 15 mm. Determine with the help of an accurate construction the

a. size of the image.
b. image distance.
c. the magnification in two ways.

Solution on page 6

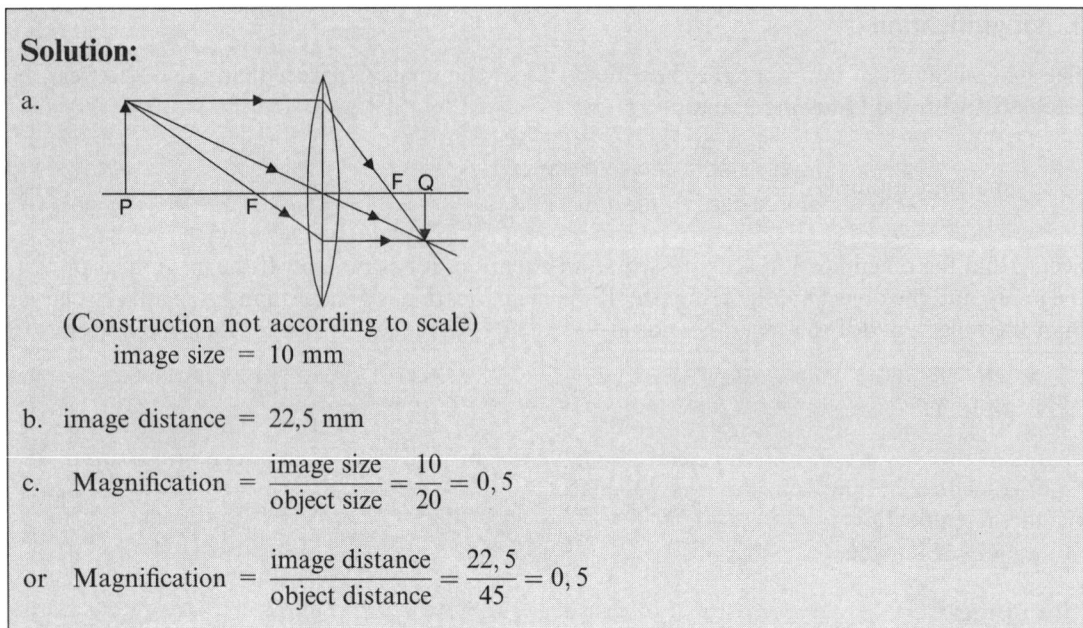

Solution:

a.

(Construction not according to scale)

image size = 10 mm

b. image distance = 22,5 mm

c. Magnification = $\dfrac{\text{image size}}{\text{object size}} = \dfrac{10}{20} = 0,5$

or Magnification = $\dfrac{\text{image distance}}{\text{object distance}} = \dfrac{22,5}{45} = 0,5$

4. The camera and the eye

There are obvious similarities between the camera lens and the eye lens. It is tabulated as follows:

EYE	CAMERA
Both are convex lenses that form reduced, inverted, real images.	
Image formed on retina	Image formed on photographic film
Scattered light is absorbed by the black layer behind the retina	Scattered light is absorbed by the black painted interior
The amount of incoming light is regulated by the iris (circular muscles)	The amount of incoming light is regulated by the diaphragm
The image is focused by accommodation – the lens changes its focal length	The image is focused by moving the lens further or closer to the film.

5. Optical instruments

Although the use of lenses in the following optical instruments also apply for Standard Grade, the **simplified ray diagrams of the film projector and the microscope only apply for Higher Grade**.

5.1 The film projector

A projector forms an enlarged, inverted, real image of the object on a screen. To accomplish this, the object (film picture) must be situated between F and 2F in front of the lens. The object must also be upside-down, so that the image can appear upright on the screen.

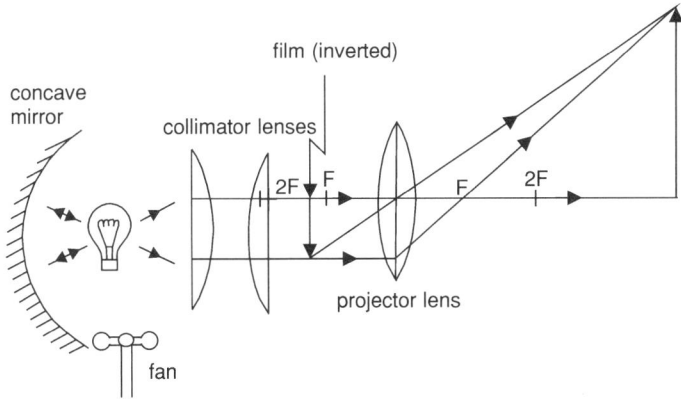

A concave mirror is placed behind the lamp to reflect all the light forward. The function of the collimator lenses is to illuminate the film equally. The powerful lamp creates much heat and therefore needs the fan to keep it cool so that it doesn't blow.

5.2 The compound microscope

The compound microscope is composed of two convex lenses with short focal lengths that are placed near to each other on the main axis. The lens closest to the object is called the objective, whereas the lens closest to the eye is called the eyepiece.

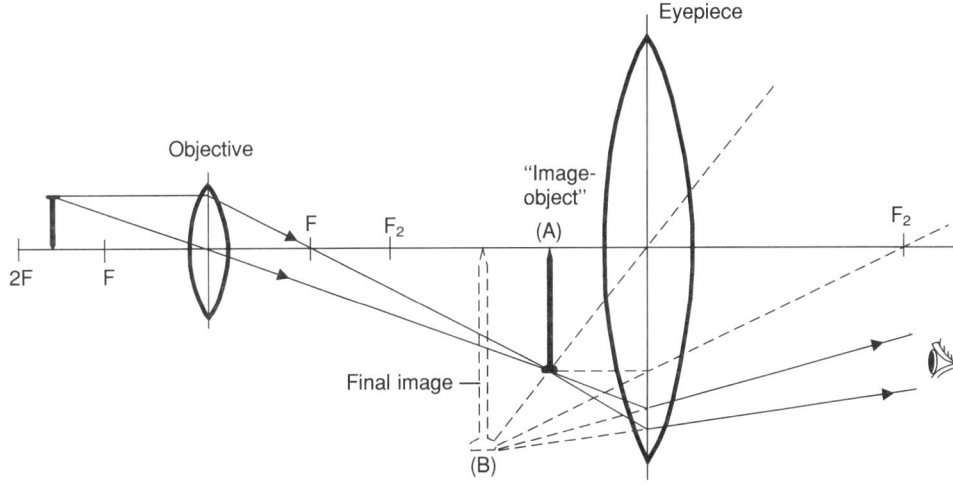

The objective is placed so that the object lies between F and 2F. A real, inverted and enlarged image (A) of the object is formed. This image now serves as an object for the second lens (eyepiece). The eyepiece is placed so that the image of the objective falls between the eyepiece and its focal point. The eyepiece therefore forms a much larger, virtual image (as with a magnifying glass) that is observed by the eye.

The "image-object" does not form really because there is no screen on which it can be projected. To determine the position of the final image, we draw **construction lines** from the end of A; one through the optical centre of the eyepiece and one parallel to the principal axis which then passes through the main focal point after refraction. The two light rays coming from the object are then produced further and leaves the eyepiece as if coming from the final image.

QUESTIONS

Section A

Different possibilities are proposed as answers to the following questions. Choose the correct answer.

1. The action of a lens depends on the principal of . . .
   A reflection.   C refraction.
   B dispersion.   D magnification.

2. Which of the following statements is false?
   A A convex lens can form real and virtual images.
   B A concave lens can only form virtual images.
   C Images formed by concave lenses, are always reduced and inverted.
   D A convex lens can form enlarged, inverted images.

3. In which of the following situations will the rays move parallel to each other after refraction?

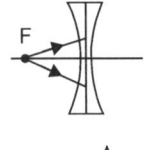 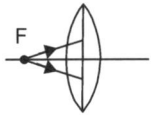 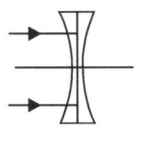 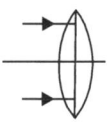

   A   B   C   D

4. In which one of the following situations will the rays diverge after refraction?

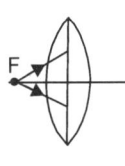 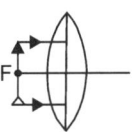 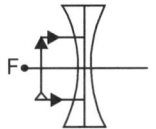 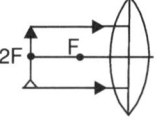

   A   B   C   D

5. No image is formed by a convex lens, if an object is placed ...

   A further than 2F   C on F.
   B on 2F.   D between F and the optical centre of the lens.

6. An object is placed between F and 2 F of a convex lens. The image that is formed, is . . .
   A reduced and upright.   C enlarged and upright
   B enlarged and inverted.   D reduced and inverted.

7. The image that is formed when an object is placed in front of a diverging lens, is . . .
   A always a reduced, virtual image.   C upright and enlarged.
   B upside down and diminished.   D always a real image and reduced.

8. The image that is formed when one looks at an object through a microscope, is . . .
   A a real image and enlarged.   C upright and enlarged.
   B inverted and reduced.   D a virtual image and enlarged.

9. The image that is formed when a projector projects it onto a screen, is . . .
   A a virtual image and enlarged.   C a virtual image and upside-down.
   B a real image and upright.   D a real image and enlarged.

10. Both the microscope and the telescope have the disadvantage that the final image is . . .
   A upright and reduced.   C enlarged.
   B inverted and reduced.   D inverted.

11. The image that is formed by one of the following objects, doesn't fit in with the others. Which one is it?
   A Camera   C Photocopier
   B Eyepiece of a microscope   D Eye

12. The retina of the human eye can be compared with a camera's . . .
   A lens.   C film.
   B diaphragm.   D lens opening.

13. The adjustable lens opening of a camera . . .
   A has the same function as the iris of the eye.
   B determines the magnification of the image.
   C forms the image on the film.
   D has the same function as the retina of the eye.

14. When the image size is 10 mm and the object size is 5 mm, the magnification is . . .
   A 2.   C 5.
   B 0,5.   D 10.

15. With a magnification of 5, the image distance of an object that is 20 mm from a lens, will be precisely . . .
   A 20 mm.   C 100 mm.
   B 5 mm.   D 4 mm.

16. When a certain apparatus is set up, it provides a magnification of 0,33. The apparatus is most likely a . . .
   A telescope.   C camera.
   B microscope.   D film projector.

17. Which of the following statements applies to the accompanying sketch?

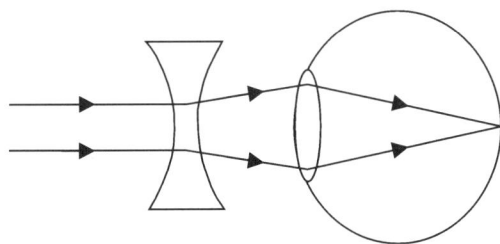

   A Far-sightedness is corrected with the help of a concave lens.
   B The concave lens replaces the eye lens's function.
   C The concave lens decreases the focal length of the eye's lens.
   D It illustrates the correction for a near-sighted eye.

Section B

1. A poor-sighted person uses a magnifying glass to read a newspaper.

1.1 Draw a ray-diagram to illustrate how the image is formed.

1.2 Give two properties of the image that is formed, except for that it is magnified.

2. An object is placed between F and 2F of a converging lens.

2.1 Draw a ray-diagram to illustrate how the image is formed.

2.2 Name three properties of the image that is formed.

2.3 What optical instrument works as in this example?

3. The photo-plate of a slide-projector is 3 m from the screen. A clear, sharp image, five times enlarged, appears on the screen.

3.1 Calculate the distance between the projector-lens and the screen (image distance).

3.2 What type of lens is used in the projector?

3.3 Between what values must the focal length of this specific lens be? Explain.

4. A real image, 20 mm high, is formed 30 mm from a convex lens. The focal length of the lens is 20 mm. Determine by means of an accurate construction

4.1 the size of the object;

4.2 how far the object is from the lens;

4.3 the magnification.

5. An object, 30 mm high, stands in front of a lens so that the image formed is 15 mm high and 50 mm behind the lens.

5.1 What type of lens is used?

5.2 Draw an accurate diagram of the rays and determine the focal length of the lens from it.

5.3 Describe the image that is formed.

5.4 What would happen to the image if the object was moved further away from the lens?

6. An object, 22 mm high, is placed 62 mm from a concave lens. The lens has a focal length of 36 mm.

6.1 Construct an accurate ray-diagram to show how the image is formed.

6.2 Measure the image size.

6.3 Calculate the magnification.

7. A concave lens forms an image 30 mm from the lens. The height of the image that is formed is 10 mm and the focal length of the lens is 50 mm.

7.1 Draw an accurate ray-diagram to illustrate how the image is formed.

7.2 Measure
(a) the height of the object
(b) the distance of the object from the lens.

7.3 Calculate the magnification of the image.

7.4 Name the properties of the image that is formed.

8. A man, 1,8 m tall, stands 20 m away from you. Your eye is capable of forming a clear image of him on your retina, 2,5 cm from your eye lens.

8.1 Draw a ray-diagram to illustrate how the image is formed.

8.2 Calculate the size of the image that is formed.

8.3 Name three properties of the image that is formed.

9. A photographer wants to enlarge a negative of 4 cm x 6 cm. The photographic enlarger is set up to obtain an image three times bigger than the negative. The negative is placed 5 cm from the lens.

9.1 What sort of lens is used in the enlarger?

9.2 Calculate how far the photographic paper must be placed to obtain a clear image.

9.3 What is the minimum size paper that can accommodate this enlargement?

10. Older people sometimes hold a book with outstretched arms.

10.1 What term would describe this eye disability?

10.2 Give a possible explanation for this phenomenon.

10.3 What type of lens is normally prescribed for this type of disability?

10.4 Make a clear sketch to illustrate where the image will be formed in the eye
(a) without the glasses;
(b) with the glasses.

11. With reference to the camera, answer the following questions:

11.1 What sort of lens is used in the camera?

11.2 What is the nature of the image that is formed?

11.3 Draw a ray-diagram to illustrate how the image is formed.

11.4 Where, on the principal axis, must the film in the camera be placed?

11.5 Why is the inside of the camera black?

11.6 Name another two optical instruments that operate on the same principle as the camera.

12. Answer the following questions regarding the compound microscope.

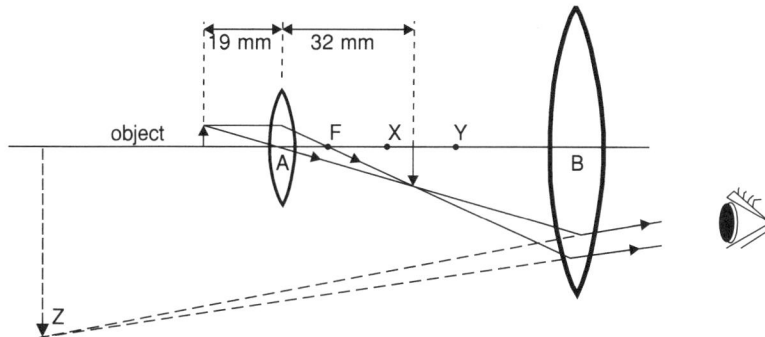

12.1 Where in terms of F and 2F is the object placed with reference to lens A?

12.2 What are lenses A and B called?

12.3 Is position X or position Y a possible focal point for the eyepiece?

12.4 Is the image, marked Z on the sketch, real or virtual?

12.5 The magnification that is obtained from the **eyepiece** is 40. Calculate the size of the image Z if the object is 0,5 mm high.

13. Redraw the following sketch to describe the principle of the compound microscope. V is the object.

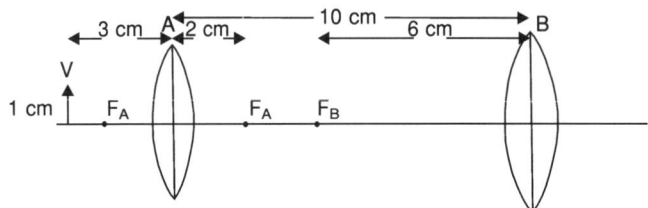

13.1 What is represented by A, B, F_A and F_B?

13.2 Complete the diagram to illustrate the image that is formed by lenses A and B.

13.3 Measure the size of the image formed by A.

13.4 Calculate the total magnification of the microscope.

14. Rewrite the following paragraph and correct all the errors that appear in it. Underline all the corrections.

The compound microscope consists of two lenses, both with long focal lengths, that are set up on two different principal axes. The objective forms a reduced, virtual image of the object, which falls between F and 2F of the eyepiece. This image then serves as the object for the eyepiece. The eyepiece then forms an enlarged, real image.

ANSWERS

Section A

1. C	**2.** C	**3.** B	**4.** C	**5.** C	**6.** B	**7.** A
8. D	**9.** D	**10.** D	**11.** B	**12.** C	**13.** A	**14.** A
15. C	**16.** C	**17.** D				

Section B

1.1

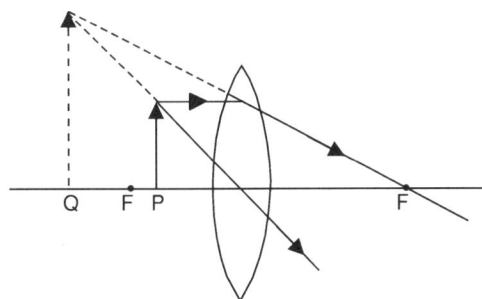

1.2 Virtual; upright

2.1

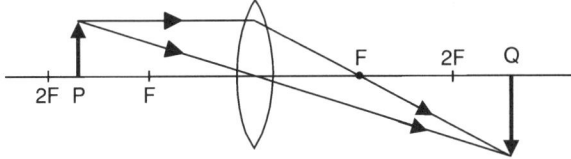

2.2 inverted
real
enlarged

2.3 film projectors, slide projectors, photographic enlargers, objective in a compound microscope

3.1

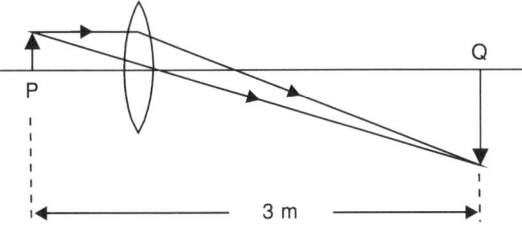

Suppose the image distance = q and the object distance = p
∴ image distance + object distance = 3   (q + p = 3) .... A

and magnification = $\dfrac{\text{image distance}}{\text{object distance}}$ = 5 $\left(\dfrac{q}{p} = 5\right)$ .... B

49

From equation B: q = 5p
Substitute in equation A:
q + p = 3   ∴ 5p + p = 3   ∴ p = 0,5 m
and q = 5p = 5 × 0,5 = 2,5 m

3.2 convex lens

3.3 to obtain an enlarged image, the object must be between F and 2F of the lens.

object distance = 0,5 m
∴ 2F must be greater than 0,5 m and F smaller than 0,5 m
∴ 2F > 0,5   and   F < 0,5
∴ F > 0,25   and   F < 0,5   ∴ 0,25 m < F < 0,5 m

4.1

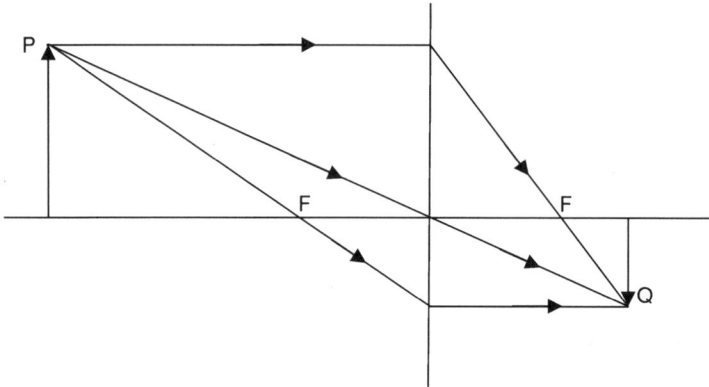

4.2 object size = 41 mm

4.3 object distance = 62 mm

4.4 magnification = $\dfrac{\text{image size}}{\text{object size}}$ OR magnification = $\dfrac{\text{image distance}}{\text{object distance}}$

$$= \frac{20}{41} = 0,49 \qquad\qquad = \frac{30}{62} = 0,48$$

5.1 convex lens

5.2

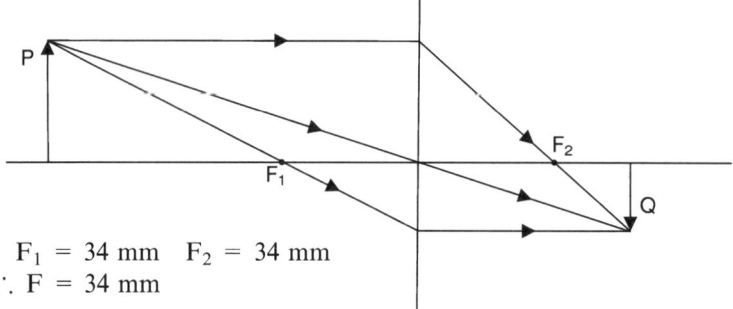

F_1 = 34 mm F_2 = 34 mm
∴ F = 34 mm

5.3 reduced; inverted; real

5.4 reduced

6.

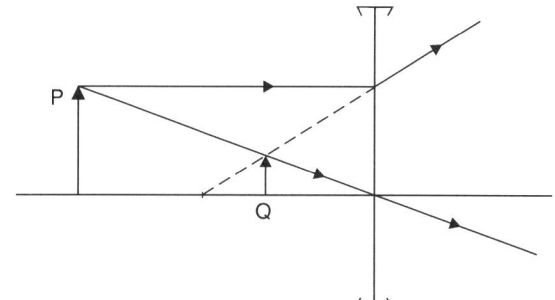

6.1 image size = 8 mm

6.2 magnification = $\dfrac{\text{image size}}{\text{object size}}$ OR magnification = $\dfrac{\text{image distance}}{\text{object distance}}$

$\qquad\qquad\quad = \dfrac{8}{22}$ $\qquad\qquad\qquad\qquad\qquad = \dfrac{22,5}{62}$

$\qquad\qquad\quad = 0,36$ $\qquad\qquad\qquad\qquad\qquad = 0,36$

7.1

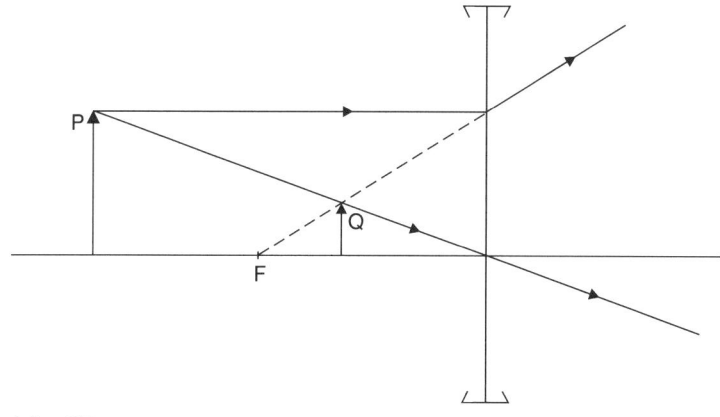

7.2 (a) 25 mm
$\quad$ (b) 76 mm

7.3 magnification = $\dfrac{\text{image distance}}{\text{object distance}} = \dfrac{76}{30} = 2,5$

$\quad$ OR

$\quad$ magnification = $\dfrac{\text{image size}}{\text{object size}} = \dfrac{25}{10} = 2,5$

7.4 reduced
$\qquad$ virtual
$\qquad$ upright

8.1

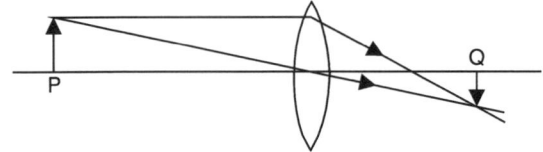

8.2 $\dfrac{\text{image size}}{\text{object size}} = \dfrac{\text{image distance}}{\text{object distance}}$

$\therefore \dfrac{\text{image size}}{1,8} = \dfrac{0,025}{2}$

$\therefore \text{image size} = \dfrac{0,025}{2} \times 1,8 = 0,0225 \text{ m}$

$= 2,25 \text{ cm}$

8.3 reduced
real
inverted

9.1 convex

9.2 $\text{magnificatioon} = \dfrac{\text{image distance}}{\text{object distance}}$

$\therefore 3 = \dfrac{\text{image distance}}{5}$

$\therefore \text{image distance} = 3 \times 5 = 15 \text{ cm}$

9.3 $3 \times (4 \text{ cm} \times 6 \text{ cm}) = 12 \text{ cm} \times 18 \text{ cm}$

10.1 far-sightedness

10.2 retina is too close to the eye-lens or the eye-lens can no longer shorten its focal length to form the image on the retina

10.3 convex lens

10.4

(a)

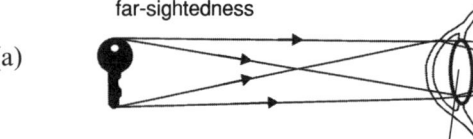

far-sightedness

eye-lens

(b)

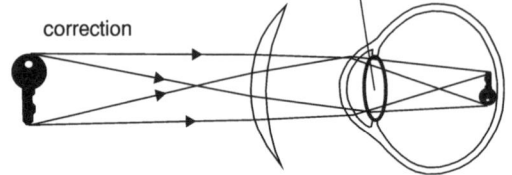

correction

11.1 convex lens

11.2 real, inverted, reduced

11.3

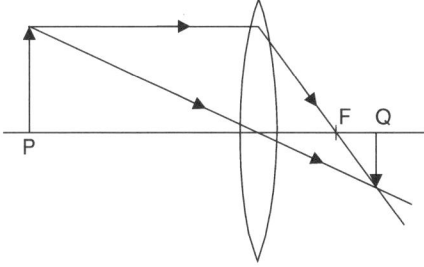

11.4 Between F and 2F on the opposite side of the lens as the object

11.5 to absorb scattered light

11.6 eye-lens; objective in a telescope

12.1 Between F and 2F

12.2 A: objective
     B: eyepiece

12.3 X

12.4 virtual

12.5 At lens A: $\dfrac{\text{image size}}{\text{object size}} = \dfrac{\text{image distance}}{\text{object distance}}$

$$\therefore \ \frac{\text{image size (A)}}{0,5} = \frac{32}{19}$$

$$\therefore \ \text{image size} = \frac{32}{19} = 0,5 = 0,84 \ \text{mm}$$

At lens B: magnification $= \dfrac{\text{image size}}{\text{object size}}$

$$\therefore \ 40 = \frac{\text{image size}}{0,84}$$

$$\therefore \ \text{image size} = 40 \times 0,84 = 33,68 \ \text{mm}$$

13.1 A: objective
     B: eyepiece
     F_A: focal point of objective
     F_B: focal point of eyepiece

13.2 (This construction is not according to scale.)

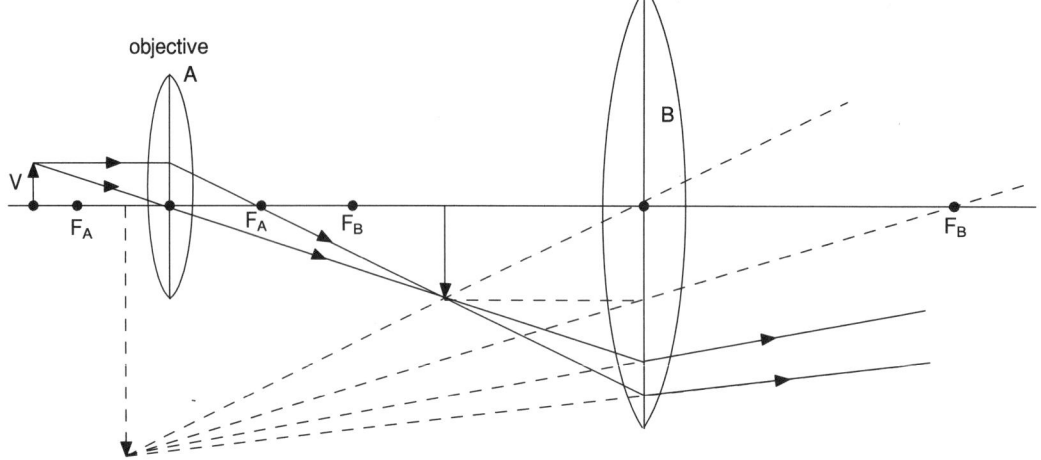

13.3 2 cm

13.4 Total magnification $= \dfrac{\text{final image size}}{\text{object size}}$

$= \dfrac{5,8}{1}$

$= 5,8$

14. The compound microscope consists of two lenses, **the first, the objective, with a short** and the second, **the eyepiece, with a long focal length mounted on the same principal axis**. The **objective forms** an **enlarged real image** of the object which is **situated further than 2F from the objective on the other side**. This image now serves as an object for the eyepiece. The eyepiece then forms an enlarged **virtual** image.

4 *Sound*

1. How is sound produced?

Energy is needed to make an object (e.g. loudspeaker) vibrate, thereby producing sound. All sounds are the result of vibrations.

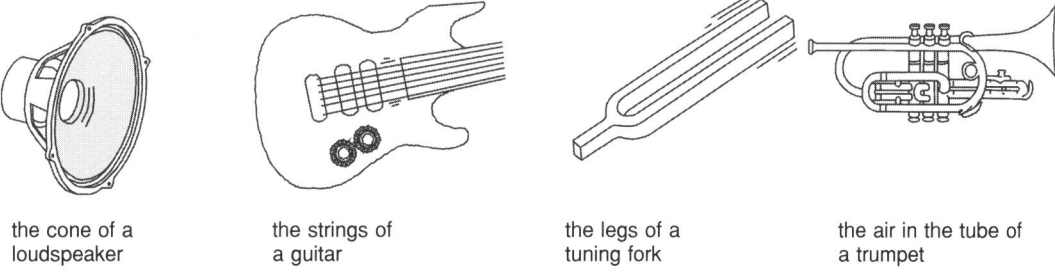

the cone of a
loudspeaker

the strings of
a guitar

the legs of a
tuning fork

the air in the tube of
a trumpet

1.1 Sound as a wave motion

The to-and-fro movements of a vibrating source of sound cause a series of compressions and rarefactions that move through a medium. Sound waves are therefore longitudinal waves.

The distance from the centre of one compression (or rarefaction) to the centre of the next compression (or rarefaction) is one wavelength.

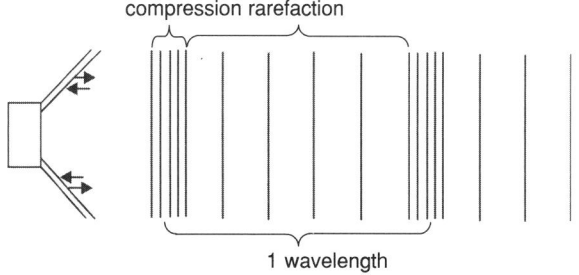

compression rarefaction

1 wavelength

Sound waves comply with the general wave equation:

Speed = frequency × wavelength   (v = f × λ)

1.2 Examples

The diaphragm of a loudspeaker vibrates at a frequency of 2 000 Hz. Calculate the wavelength of the sound wave if it moves at a speed of 340 m.s^{-1} through the air.

Solution:

f = 2 000 Hz;   v = 340 m.s^{-1}

$$\lambda = \frac{v}{f} = \frac{340}{2\,000} = 0,17 \text{ m}$$

Example 2:

Calculate the wavelength of the sound waves that are produced by a tuning fork marked: C -264 Hz. Accept the speed of the sound waves as 340 m.s^{-1}.

Solution:

f = 264 Hz;   v = 340 m.s^{-1}

$$\lambda = \frac{v}{f} = \frac{340}{264} = 1,29 \text{ m}$$

Example 3:

Calculate the frequency with which a guitar string vibrates to produce a sound wave with a wavelength of 0,8 m. Accept the speed of the sound wave as 330 m.s^{-1}

Solution:

v = 330 m.s^{-1}; λ = 0,8 m

$$f = \frac{v}{\lambda} = \frac{330}{0,8} = 412,5 \text{ Hz}$$

1.3 The transmission of sound

Sound waves need a material medium in order to be propagated. It can therefore not move through a vacuum; as light waves can do.

Sound travels best in a solid, very well through a fluid and the worst through a gas. Sounds that move through a solid or fluid, are therefore heard much further and faster than sounds that move through gases.

Sound also travels faster through a medium with a higher temperature. The speed of sound through air is 331 m.s^{-1} at 0 °C and 343 m.s^{-1} at 20 °C.

2. Characteristics of sound

2.1 Reflection of sound waves

Sound waves, as all other wave motions, are reflected according to the reflection laws. When waves strike an obstruction, they are reflected so that the **angle of incidence ($\angle$ i) is equal to the angle of reflection ($\angle$ r)**.

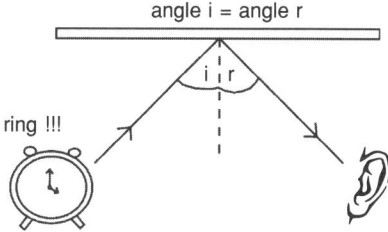

When a person claps his hand a certain distance from a wall, the sound wave is reflected by the wall back to the person, and is heard as a **echo**. The time interval between the original sound and the echo is the time that the sound wave takes to travel from the person to the reflective surface and back again.

In a hall with smooth walls, the voice of a speaker is reflected repeatedly between the walls. The **multiple reflections** or **reverberations** don't reach the observer's ear simultaneously so that it is difficult to understand what the speaker is saying. The **acoustics** of such a hall is poor. Curtains can be hung up on the walls to absorb the reflected noise and therefore improve the acoustics.

The echo of sounds has many practical uses.

Parabolic reflectors reflect all sound to a focus point to thereby intensify the sound.

Bats send out ultra-sonic sounds that reflect off insects and thereby can track down their prey.

Sonar is used to search for submarines or to determine the depth of the seabed.

Example 4:

An echo is heard 4 seconds after a sound is created. If the sound moves at 330 m.s^{-1}, calculate the distance between the source of sound and the reflective surface.

Solution:

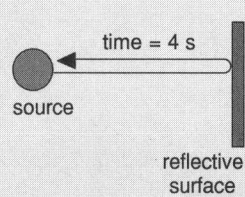

$$\text{time} = 4 \text{ s}; \quad \text{speed} = 330 \text{ m.s}^{-1}$$

$$\text{distance} = \text{speed} \times \text{time}$$
$$= 330 \times 4 = 1\,320 \text{ m}$$

$\therefore$ distance that sound moves $= 1\,320$ m
(i.e. the distance from the source to the reflective surface and back to the source $= 1\,320$ m)
$\therefore$ distance between source and reflective surface
$$= \frac{1\,320}{2} = 660 \text{ m}$$

Example 5:

The speed of sound in sea water is 1 500 m.s⁻¹. A submarine sends out impulses that are received after 20 s. Calculate the depth of the sea.

Solution:

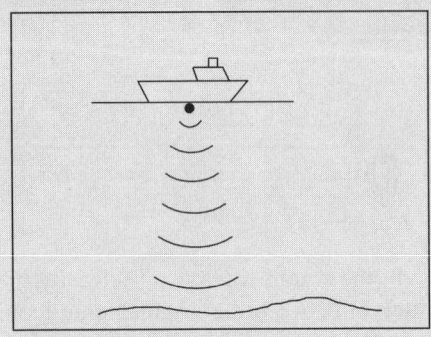

speed = 1 500 m.s^{-1};   time = 20 s

$\therefore$ distance = speed × time
= 1 500 × 20 = 30 000 m

$\therefore$ depth = $\dfrac{30\,000}{2}$ = 15 000 m

Example 6:

A boy stands between two high buildings. When he claps his hands together, he hears the echo from the one building after 2 seconds and the echo from the other building after 3 seconds. Calculate the distance that the buildings are from each other. Accept the speed of sound as 330 m.s⁻¹.

Solution:

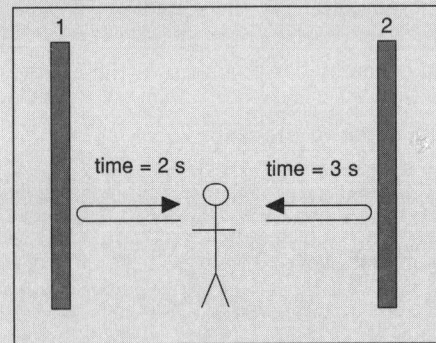

For building 1:
distance = speed × time
= 330 × 2 = 660 m

$\therefore$ distance between boy and building 1:

distance = $\dfrac{660}{2}$ = 330 m

For building 2
distance = speed × time
= 330 × 3 = 990 m

$\therefore$ distance between boy and building 2:

distance = $\dfrac{990}{2}$ = 495 m

Distance between the two buildings is therefore
330 + 495 = 825 m

2.3 Refraction of sound waves

Sound waves are refracted if they move obliquely from one medium to another. The speed at which the waves move, changes, so that the wave direction also changes.

Sound waves move faster in warm air than cold air. At night, when the air layers closer to the earth's surface are colder, sounds are heard better because the sound waves are refracted back to the earth. The sound waves are constantly bent away from the normal (as in the sketch).

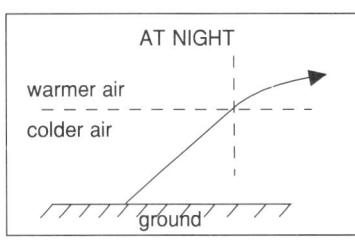

By day, the air layers near the earth's surface are hotter, so that sounds are refracted away from the earth's surface and are not heard so well. The sound waves are constantly bent closer to the normal (as in the sketch).

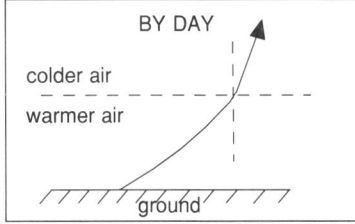

2.4 Interference of sound waves

Sound waves undergo interference when two waves meet each other. When two compressions or two rarefactions of longitudinal waves meet, constructive interference takes place, so that a louder sound is heard. When the compression of one wave meets the rarefaction of another wave, destructive interference takes place, so that a softer or no sound is heard.

2.5 Resonance

Each object has a natural frequency at which it vibrates. When a tuning fork is knocked on a table, it vibrates at its natural frequency e.g. 512 Hz or 256 Hz. These vibrations of the tuning fork can be intensified when another tuning fork that vibrates at the same frequency, is held close to the first one.

> When a tuning fork that vibrates at the natural frequency of a second tuning fork, is held close to the second tuning fork, the second tuning fork will also begin to vibrate. This phenomenon, where a **system begins to vibrate at its natural frequency due to the impulses it receives from another source that is vibrating at the same frequency,** is known as **resonance**.

3. Sound and the ear

When sound waves enter the human ear, the compressions and rarefactions of the wave cause the tympanic membrane to vibrate. These vibrations are intensified by three tympanic bones and transferred to the auditory nerve. In the brain the impulses are then interpreted as a certain sound.

The human ear hears sounds between more-or-less 20 Hz and 20 000 Hz. Sounds with a frequency higher than 20 000 Hz can therefore not be heard by the human ear and are known as **ultra-sonic** or **supersonic**.

Sound waves can be illustrated as a graph by an **oscilloscope**.

3.1 Pitch of sounds

The pitch of a sound is how high or low the note sounds. The pitch of a note depends on the frequency of the vibration. An increase of the frequency causes the pitch to rise. The wavelength of the vibration decreases accordingly whereas the amplitude remains the same.

The diagram illustrates the oscilloscope pattern of a note of which the volume stays the same, but has a higher pitch at (a) compared to a lower pitch at (b).

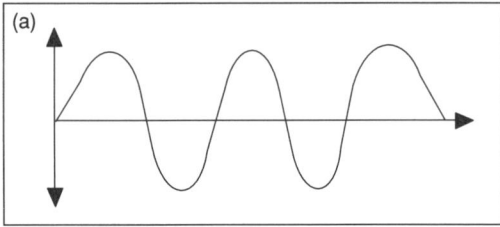
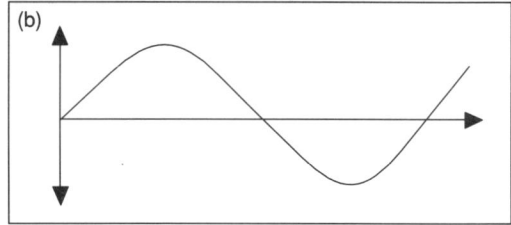

3.2 Loudness of sounds

The volume of a note depends on the amplitude of the vibration. An increase in the amplitude causes the volume to increase. The wavelength and the frequency of the vibration stay the same.

The diagram illustrates the oscilloscope pattern of a note of which the pitch remains the same, but is soft in (a) and loud in (b).

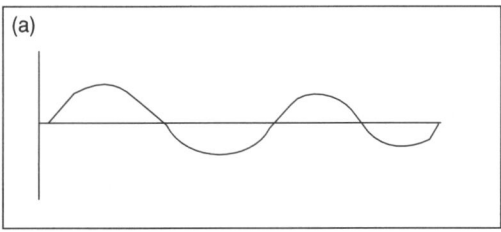
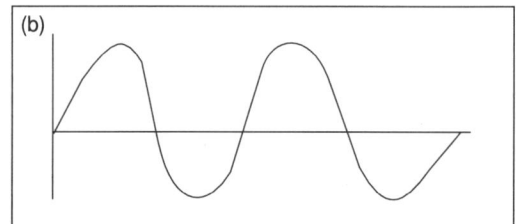

3.3 Noise and musical sounds

Regular vibrations cause musical notes. The wave form of such a note remains the same and is repeated over and over. A noise is the consequence of irregular vibrations that have no fixed pattern.

The diagrams illustrate the oscilloscope patterns of certain musical notes as well as a noise.

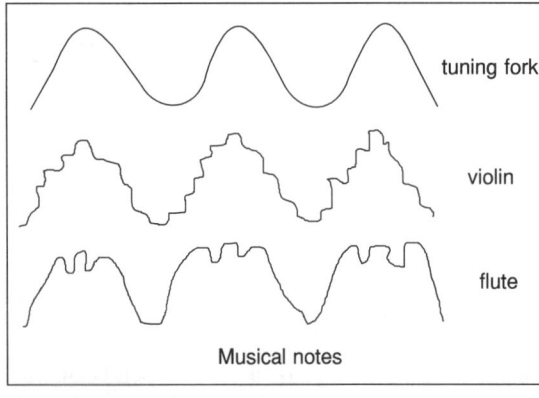
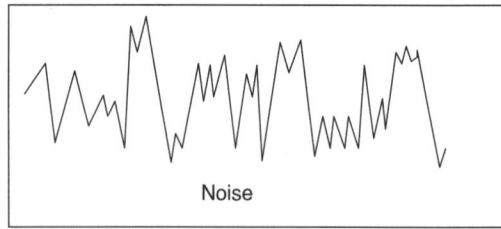

QUESTIONS

Section A

Different possibilities are proposed as answers to the following questions. Choose the correct answer.

1. Sound waves are longitudinal waves because . . .
   A they cannot move through a vacuum.
   B they can be transmitted by a material medium.
   C the vibration of the particles is perpendicular to the direction of the wave.
   D the vibration of the particles is in the same direction as that in which the wave is moving.

2. Which of the following are longitudinal waves?
   A Water waves
   B Light waves
   C Sound waves
   D Radio waves

3. Sound will be best transmitted by . . .
   A air.
   B water.
   C iron.
   D wood.

4. Sound will **not** be transmitted by . . .
   A water.
   B air.
   C a vacuum.
   D iron.

5. Which of the following is **not** an example of a regular repetitive motion?
   A A guitar string that vibrates
   B The wings of a bee that moves quickly up and down
   C A metre stick that swings to-and-fro
   D The keys of a typewriter as you type normally

6. When a tuning fork is hit with a hammer, a sound wave is produced . . .
   A with a constant amplitude.
   B of which the wavelenghts gradually decrease.
   C of which the frequency gradually decreases.
   D of which the amplitude gradually decreases.

7. The speed of a sound wave is . . .
   A the same in all mediums in which it moves.
   B the highest in a solid medium.
   C the highest in a gas medium.
   D the lowest in a fluid medium.

8. The wavelength of sound is . . .
   A the same in all media.
   B the longest in gases.
   C the shortest in fluids.
   D the longest in solids.

9. A tuning fork is marked 256 Hz. The speed of sound in air is 330 m.s^{-1}. The wavelength of the note that is produced by the tuning fork is . . .
   A 1,29 m.
   B 0,78 m.
   C 84,48 km.
   D $1,29 \times 10^{-3}$ m.

10. Sound waves . . .
A consist of successive compressions and rarefactions in the medium.
B consist of successive sound crests and sound troughs.
C are transmitted through a vacuum.
D move at the same speed through all substances.

11. Sound is transmitted more slowly through . . .
A metals than through water.
B water than through gases.
C cool gases than through warm gases.
D air than through carbon dioxide.

12. An echo is heard 4 s after a sound has originated. If the sound moves at 330 m.s^{-1}, the distance between the sound source and the reflecting surface is . . .
A 1 320 m.
B 660 m.
C 82,5 m.
D 165 m.

13. The speed of sound in sea water is 1 500 m.s^{-1}. A sonar apparatus sends out signals which are received after 20 s. The depth of the sea at that point is approximately . . .
A 1 500 m.
B 30 000 m.
C 150 m.
D 15 000 m.

14. The phenomenon that sound can be heard over a greater distance in the morning than in the afternoon, is explained by the fact that . . .
A the speed of sound depends on the temperature of the air.
B there is less noise in the morning.
C sound is a longitudinal wave motion.
D sound and light do not travel at the same speed.

15. It takes sound approximately 3 s to travel through a distance of 1 km. If you hear the thunder 9 seconds after having seen the lightning flash, the distance between you and the lightning flash is . . .
A 27 km.
B 9 km.
C 3 km.
D $\frac{1}{3}$ km.

16. Which of the following statements is correct?
A High notes move at the same speed through glass than low notes.
B High notes move faster through glass than low notes.
C Low notes move faster through air than high notes.
D High notes move faster through air than low notes.

17. Which of the following statements for sound and light is correct?
A The speed of both is 340 m.s^{-1}.
B Both can be reflected.
C Both can move through a vacuum.
D Both can move through iron.

18. A moderate increase in the amplitude of a vibration will . . .
A lower both the loudness and the pitch.
B raise both the loudness and the pitch.
C lower the loudness and raise the pitch.
D raise the loudness but not the pitch.

19. Two notes played on different musical instruments, always differ in . . .
A speed.
B wavelength.
C wave form.
D amplitude.

20. The sound produced by a tuning fork becomes softer after a while, because the . . .
A amplitude decreases.
B frequency decreases.
C frequency increases.
D amplitude increases.

21. Which of the following statements is correct?
  I The higher the frequency the higher the pitch.
  II The higher the frequency, the shorter the wavelength.
  III The higher the frequency, the faster the sound moves.

A Only I   C Only I and II
B Only II   D Only I and III

22.

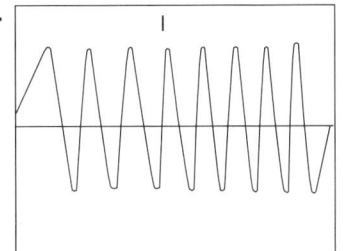 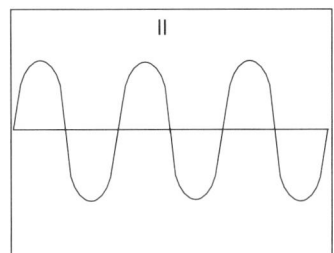 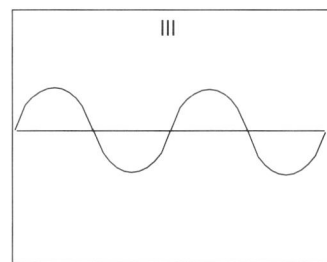

I, II and III represent sound waves on the screen of an oscilloscope. Which of the statements is correct?

A I and II have the same frequency.   C II and III's pitch is greater than that of I.
B II and III have the same loudness.   D I and II have the same loudness.

Section B

1.1   What is necessary for sound waves to be transmitted from one place to another?

1.2   Can sound waves move through a vacuum? Explain your answer.

2.   The speed of sound in air is 330 m.s^{-1}. Calculate the frequency of a note with a wavelength of 2 m.

3.   The frequency of a note is 50 Hz. What does it mean?

4.   From a vertical cliff an echo of your voice is heard after 3 s. How far from the cliff are you standing? (Speed of sound through air is 330 m.s^{-1}).

5.   The characteristics of sound waves are examined.

5.1   Give, in the form of a table, three differences between light and sound waves.

5.2   Explain why echoes cannot be heard in a small room.

6.   The time-keeper at a 100 m athletics race presses his stopwatch when he hears the shot of the starter's pistol.

6.1   Explain why it is an inaccurate method and suggest a better hand-method.

6.2   Calculate by how many seconds the time-reading will differ from the correct reading, if the speed of sound is taken to be 340 m.s^{-1}.

7.   A girl standing between two cliffs, discovers that when she claps her hands, she can hear the echo from the one cliff after 1,5 s and from the other cliff after 2 s. Calculate the distance between the two cliffs if the speed of sound in air can be taken as 340 m.s^{-1}.

8.1 A string vibrates with a period of 0,002 s. Calculate the wavelength of the note that is produced if the speed of sound in air is 340 m.s⁻¹.

8.2 What will happen to the

(a) frequency

(b) speed

(c) wavelength when the sound waves enter a steel object from air?

9. A ship travels 850 m away from a perpendicular cliff. The sirens sound and the echo is heard after 5 seconds. Calculate the speed of sound from what is given.

10. A boy standing 85 m from a high wall, realises that the echo coincides with the stroke if he hits a drum every 0,5 s.

10.1 Calculate the speed of the sound wave from this information.

10.2 If the drum's membrane vibrates at 100 Hz, calculate the wavelength of the sound waves.

11. In an experiment to determine the speed of sound, a glass tube which is gradually immersed in a bowl of water is used. A tuning fork with a frequency of 256 Hz is vibrated above the glass tube. It is determined that the distance between two resonance crests of the sound waves is 67 cm. Calculate the speed of sound if it is given that the distance between two resonance crests is equal to half a wavelength.

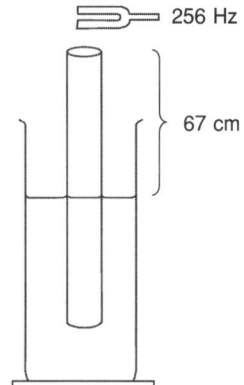

12. In an experiment to determine the speed of sound, two boys stand 450 m from each other. The one boy shoots into the air with a revolver, while the other boy presses a stopwatch as soon as he sees the smoke from the revolver. The moment that he hears the shot, he presses the stopwatch for the second time. He notices that the time lapse is 1,32 s. Calculate the speed of sound waves from the above information.

13. Dolphins communicate with each other by means of whistling noises. The frequency of these sound waves is 300 Hz. The speed of sound in sea water at 20 °C is 1 480 m.s⁻¹.

13.1 Calculate the wavelength of the sound waves (answer to two decimal units).

13.2 Two schools of dolphins communicate with each other. It takes the whistling noises of one school 1,5 seconds to reach the other school. How far are the two schools of dolphins from each other?

14. Two boys, Jan and Shaun, stand 880 m from each other next to a railway track, as shown in the sketch. Jan puts his ear against the rail, while Shaun hits the track with a stone.

A girl, Lindi, is curious about what the boys are doing, and watches from point B in the sketch.

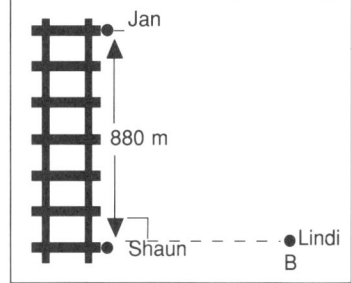

14.1 Calculate how long it takes before Jan hears the sound once Shaun has hit the track. Take the speed of sound through iron as 5 280 m.s^{-1}.

14.2 Calculate how far Lindi must stand from Shaun to hear the sound at precisely the same time as Jan, if the speed of sound through air can be taken as 340 m.s^{-1}.

15. A tuning fork is hit with a hammer and allowed to vibrate for a while. How do the following change?

15.1 The speed of the sound wave

15.2 The frequency of the sound wave

15.3 The amplitude of the sound wave

15.4 The loudness of the sound

15.5 The pitch of the sound during this period

15.6 Make a neat diagram to illustrate this sound wave.

16. A series of sound waves of a pure note is observed on a ossciloscope-screen during an experiment. Illustrate, by means of simple diagrams with lables, how the observed pattern changes, if . . .

16.1 the loudness of the note decreases.

16.2 the pitch of the note increases.

16.3 the pure note changes to a noise.

17. A sound gradually becomes louder. How does the

17.1 wavelength

17.2 frequency

17.3 amplitude and

17.4 speed of this sound wave change?

18. Radio Good Hope broadcasts at a frequency of 88,2 MHz (megahertz). Two loudspeakers of Radio Good Hope are set up at an open-air festival terrain, as shown in the sketch. While you move around in the area (marked with the block), you find that the sound is louder at certain places than others.

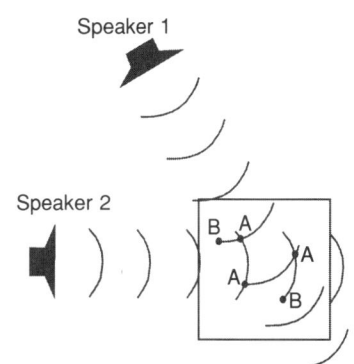

18.1   What sound phenomenon takes place in the area of the block?

18.2   Will the sound be louder or softer at the points marked A? Explain your answer.

19.

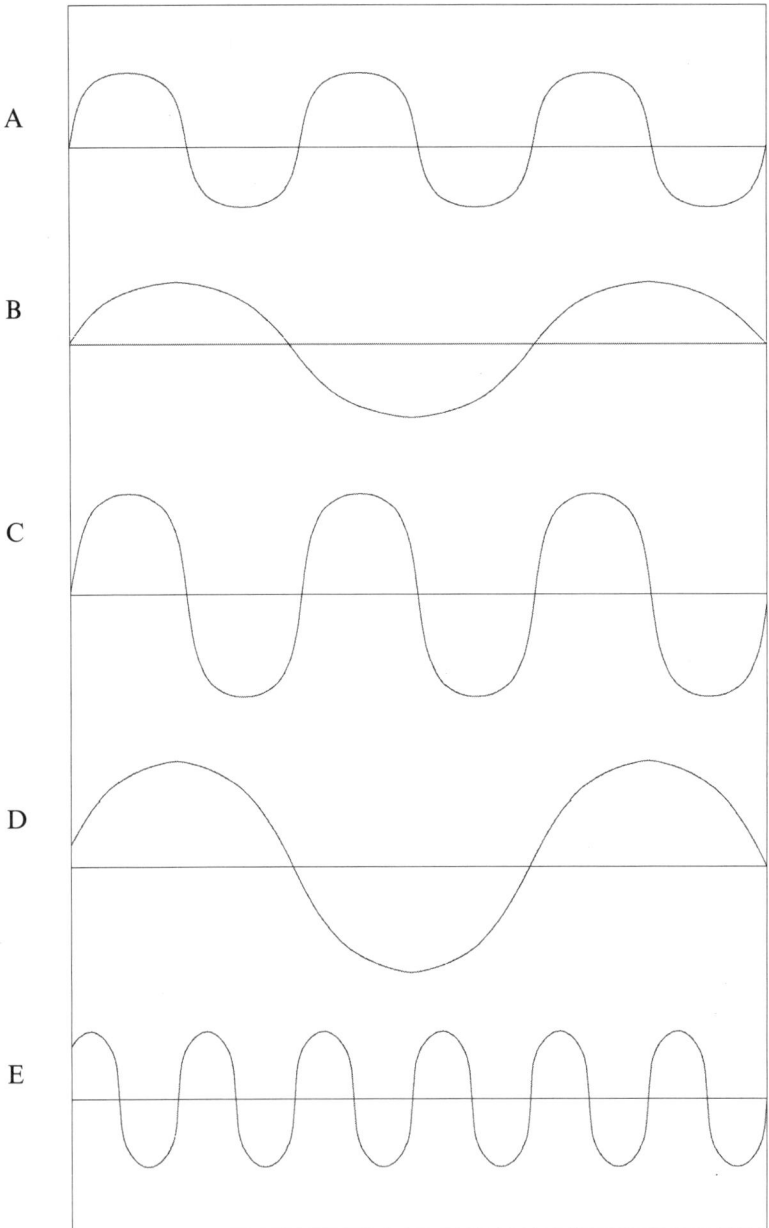

Which of the above wave patterns illustrate wave patterns that

19.1 are of the same volume?

19.2 have the same pitch?

19.3 is the highest note?

19.4 is the lowest note?

ANSWERS

Section A

1. D	**2.** C	**3.** C	**4.** C	**5.** D	**6.** D	**7.** B
8. D	**9.** A	**10.** A	**11.** C	**12.** B	**13.** D	**14.** A
15. C	**16.** A	**17.** B	**18.** D	**19.** C	**20.** A	**21.** C
22. D						

Section B

1.1 a medium

1.2 No, the sound waves need a medium that can vibrate.

2. $f = \dfrac{v}{\lambda} = \dfrac{330}{2} = 165$ Hz

3. The waves repeat 50 times per second.

4. distance = speed × time = 330 × 3 = 990 m

∴ distance from cliff = $\dfrac{990}{2}$ = 495 m

5.1

Sound	Light
• speed of 340 m.s^{-1}	speed of 3×10^8 m.s^{-1}
• longitudinal waves	transverse waves
• needs a medium in which to travel	can travel through a vacuum

5.2 The distance that the sound travels is very short so that the echo is heard virtually simultaneously with the original sound.

6.1 The sound wave's speed is relatively slow; the time-keeper must press his stopwatch when he sees the pistol's smoke – light waves travel much faster.

6.2 speed = $\dfrac{distance}{time}$ ∴ time = $\dfrac{distance}{speed}$

$= \dfrac{100}{340} = 0,294$ s

7. cliff 1: distance = speed × time = $340 \times \dfrac{1,5}{2}$ = 255 m

cliff 2: distance = speed × time = $340 \times \dfrac{2}{2}$ = 340 m

distance between cliffs = 255 + 340 = 595 m

8.1 $T = 0,002$ s

$\therefore f = \dfrac{1}{0,002} = 500$ Hz

$\therefore \lambda = \dfrac{v}{f} = \dfrac{340}{500} = 0,68$ m

8.2.1 frequency stays the same

8.2.2 speed increases

8.2.3 wavelength increases.

---

9. speed $= \dfrac{\text{distance}}{\text{time}} = \dfrac{850 \times 2}{5} = 340$ m.s^{-1}

10.1 speed $= \dfrac{\text{distance}}{\text{time}} = \dfrac{85 \times 2}{0,5} = 340$ m.s^{-1}

10.2 $\lambda = \dfrac{v}{f} = \dfrac{340}{100} = 3,4$ m

---

11. $\lambda = 0,67 \times 2 = 1,34$ m   f = 256 Hz

$\therefore v = f\lambda = 256 \times 1,34 = 343$ m.s^{-1}

---

12. speed $= \dfrac{\text{distance}}{\text{time}} = \dfrac{450}{1,32} = 340,9$ m.s^{-1}

13.1 $\lambda = \dfrac{v}{f} = \dfrac{1\,480}{300} = 4,93$ m

13.2 distance = speed × time = $1\,480 \times 1,5 = 2\,220$ m

---

14.1 time $= \dfrac{\text{distance}}{\text{speed}} = \dfrac{880}{5\,280} = 0,167$ s

14.2 distance = speed × time = $340 \times 0,167 = 56,78$ m

15.1	stays the same
15.2	stays the same
15.3	decreases
15.4	decreases
15.5	stays the same
15.6	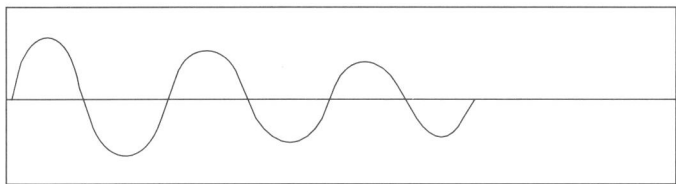

16.1	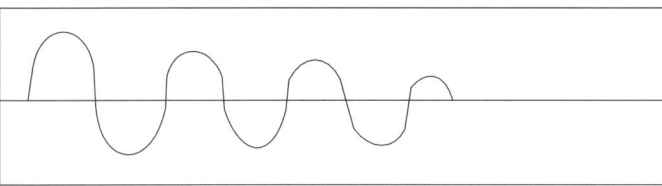
16.2	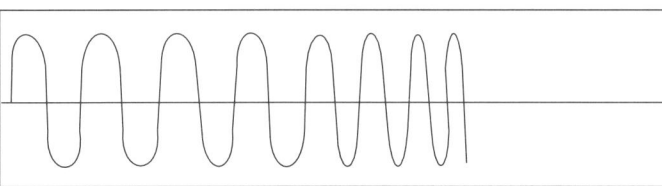
16.3	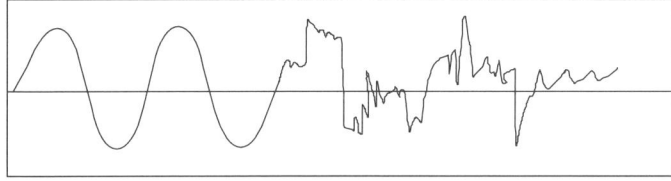

17.1	stays the same;
17.2	stays the same;
17.3	increases;
17.4	stays the same.

18.1	interference
18.2	louder – two waves fronts (crests) meet and constructive interference takes place.

19.1	A, B, E and C, D
19.2	A, C and B, D
19.3	E
19.4	B, D.

5 *Electrical Quantities*

1. Revision of Grade 9 work

1.1 Current convention

The direction of conventional current is from the positive pole of the cell, through the circuit to the negative pole of the cell.

1.2 Cells connected in series and parallel

The figure shows how cells are connected in series and parallel:

In series: In parallel:

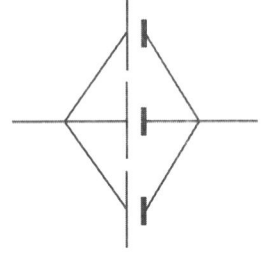

The more cells connected in series in a circuit, the bigger is the current produced and the greater is the ability of the battery to supply current (the emf of the battery).

The more cells connected in parallel in a circuit, the longer the cells can supply the same current, but they cannot supply a larger current. The maximum potential of the battery is still the same as for a single cell.

2. Electric current

An electric current consists of a flow of charge (electrons in the case of metals and ions in the case of liquids).

The unit in which electric charge is measured, is the **coulomb (C).**

Electric current is defined as the **amount of charge which flows past a point per second.**

$$\text{Current (in ampère)} = \frac{\text{Charge (in coulomb)}}{\text{Time (in seconds)}}$$

In symbols: $I = \dfrac{Q}{t}$

Electric current is measured in **ampère (A)** with an **ammeter** which is connected **in series** in a circuit.

Note: The formula Q = It can be used to define **one coulomb of charge**:

> **One coulomb of charge passes any cross section of a conductor in one second, when a current of one ampère flows through the conductor.**

3. Potential Difference

In the accompanying circuit a charge + Q has a higher potential energy value at point A than at point B. There is, therefore, a difference in potential energy, or a **potential difference**, between points A and B, which is equal to the work done to move the charge between points A and B.

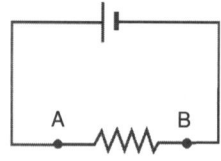

Definition:

> The **potential difference** between two points in a conductor is the work done per unit charge to move a positive charge from one point to another.

$$\text{Potential difference (in volts)} = \frac{\text{Work done (in joules)}}{\text{Charge (in coulomb)}}$$

In symbols: $V = \dfrac{W}{Q}$

Potential difference is measured in **volts (V)** with a **voltmeter** which is connected in **parallel** in a circuit.

Note: A current can therefore only flow between two points in a conductor if a potential difference between the points exists.

4. Resistance

If the number of cells in the accompanying circuit is increased (e.g. from one to five), while the temperature of the resistor is kept constant, it is found that the voltmeter reading as well as the ammeter reading increase.

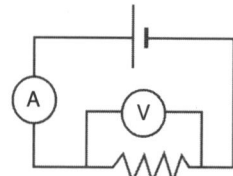

A graph of the potential difference versus the current forms a straight line. The gradient of the graph, the ratio V/I, remains the same for a specific resistor and is called the resistance of that resistor.

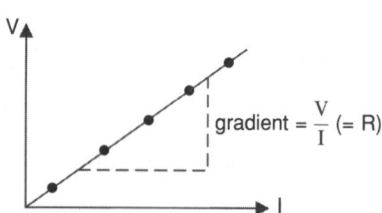

The relationship is summed up in **Ohm's Law:**

> The current between any two points in a conductor is directly proportional to the potential difference between these points, provided that the resistance of the conductor remains constant (i.e. provided the temperature of the conductor remains constant).

$$\text{Resistance (in ohm)} = \frac{\text{potential difference (in volts)}}{\text{Current (in ampere)}}$$

In symbols: $R = \dfrac{V}{I}$

The **unit of resistance** is **ohm** (Ω) and is **defined** as follows:

> **A conductor has a resistance of one ohm if a current of one ampere passes through it when a potential difference of one volt is maintained between its ends.**

Factors which determine the resistance of a conductor, are the

- **type of material** used – e.g. nichrome has a higher resistance than copper.
- **length** of the conductor – the longer the conductor, the greater its resistance.
- **thickness** of the conductor – the thicher the conductor, the smaller its resistance.
- **temperature** of the conductor – the hotter the conductor, the greater its resistance.

5. Resistors connected in series and parallel

5.1 Resistors connected in series

- $I_1 = I_2 = $ constant (in the diagram: $A_1 = A_2 = $ constant)
- $V_T = V_1 + V_2 + V_3$
  resistors connected in parallel
  are potential dividers
- $R_{\text{totaal}} = R_1 + R_2 + R_3$
- the more resistors in series, the greater
  the resistance and the smaller the current

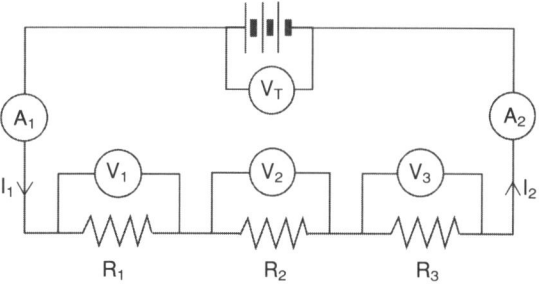

5.2 Resistors connected in parallel

- $V_1 = V_2 = V_3 = V_4 = V$ (constant)

- $I_{\text{total}} = I_1 + I_2 + I_3$

  Resistors connected in parallel are current
  dividers.

- $\dfrac{1}{R_{\text{total}}} = \dfrac{1}{R_1} + \dfrac{1}{R_2} + \dfrac{1}{R_3}$

- the more resistors in parallel, the smaller the
  resistance and the greater the current.

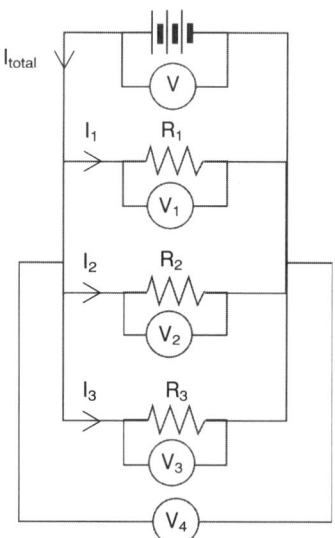

6. EMF (E), "lost volts" (V_1) and internal resistance (r)

Emf is the **maximum amount of energy per unit charge** which the cell can produce. It is measured with voltmeter V when no current is flowing through the circuit. The emf (E) of a cell cannot change.

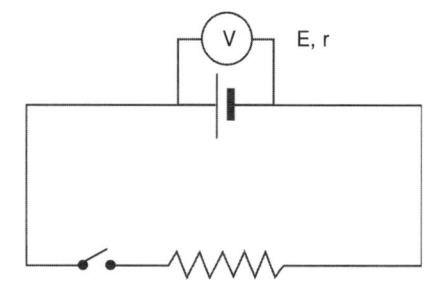

When a current flows through the circuit, most of the energy is transferred to the resistor, but a small amount is lost inside the cell due to the resistance experienced by the moving charges when they collide with the atoms inside the cell. This resistance of the cell is known as the **internal resistance** (r) of the cell.

This energy which is lost inside the cell (it is actually transferred to heat), is known as the **"lost volts"**. The **"lost volts"** is the **difference between the emf of the cell and the voltmeter reading** when a **current is flowing.**

The emf therefore is the total of the potential difference (V) of the circuit and the potential difference of the cell (the "lost volts").

$$E = V + \text{"lost volts"} = IR + Ir = I(R + r)$$

Note: The calculation of "lost volts" and internal resistance is NOT in the Grade 10 core syllabus.

7. Calculations

7.1 Summary of formulae

By using the following formulae, **all** current electricity problems can be solved.

$$R = \frac{V}{I}$$

$$r = \frac{\text{"lost volts"}}{I}$$

$$Q = It$$

$$V = \frac{W}{Q}$$

R: resistance
r: internal resistance
V: potential difference
I: current
W: work done or energy converted
t: time in seconds
Q: charge

7.2 Hints with the solving of problems

- If the circuit appears "strange", try to redraw it until it looks like a diagram which you are used to.
- Components are in parallel when you can recognize **a branching** in the circuit.
- Values substituted into formulae must "fit", e.g. $R = V/I$; the V and I values must be measured over **the same** component.
- The following do not change in a circuit: The **emf** of a cell and the **resistance** of a **resistor**.
- Hints for the actual solving of a problem:
    1. Write the given information in symbolic form, e.g. $I = 2$ A.
    2. In questions with subsections, the answer of the previous subsection usually is needed in the following question.

Note to Learners and Teachers:

The syllabus makes provision for calculations based on circuit diagrams where only one resistor occurs. The syllabi of certain provincial education departments, however, make provision for an extension in the number of resistors, but specify that it may not be used in examinations for promotion purposes. The authors of this book though are convinced that many teachers do include these types of questions in examinations, and have therefore included a number of comprehensive problems in this regard.

Example 1:

Calculate the total resistance of the following:

(a)
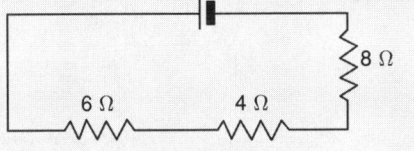

(b)
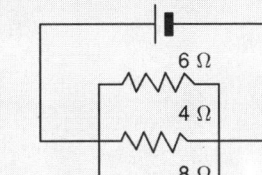

(c)
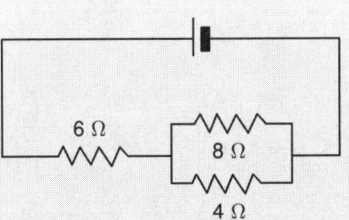

(d)
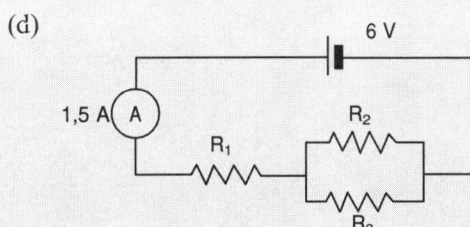

Solution:

(a) $R_{tot} = 6 + 4 + 8 = 18 \ \Omega$

(b) $\dfrac{1}{R_{tot}} = \dfrac{1}{6} + \dfrac{1}{4} + \dfrac{1}{8} = \dfrac{4 + 6 + 3}{24} = \dfrac{13}{24}$

$$\therefore R_{tot} = \dfrac{24}{13} = 1,85 \ \Omega$$

(c) $\dfrac{1}{R_{\parallel}} = \dfrac{1}{8} + \dfrac{1}{4} = \dfrac{1 + 2}{8} = \dfrac{3}{8} \quad \therefore R_{\parallel} = \dfrac{8}{3} = 2,67 \ \Omega$

$$\therefore R_{tot} = 6 + 2,67 = 8,67 \ \Omega$$

(d) $R_{tot} = \dfrac{V_{tot}}{I_{tot}} = \dfrac{6}{1,5} = 4 \ \Omega$

75

Example 2:

Two cells are connected in series with two resistors R_1 and R_2, together with an ammeter and a switch. A voltmeter connected over R_1, gives a reading of 4 V. The resistance of R_2 is 3 Ω and the reading on the ammeter is 1,5 A.

(a) Draw the circuit diagram
(b) Calculate the resistance of R_1
(c) Calculate the potential difference over R_2
(d) Calculate the emf of each cell
(e) Calculate the amount of charge flowing through the ammeter in two minutes.
(f) Calculate the amount of energy converted by R_2 into light and heat in two minutes.

Solution:

(a)

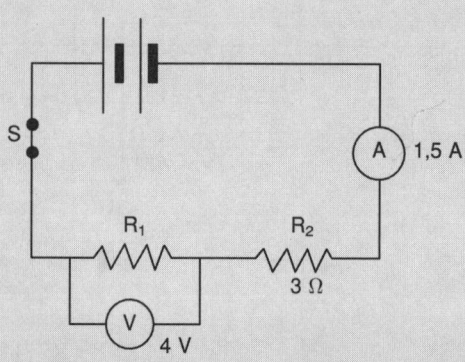

(b) $R_1 = \dfrac{V}{I} = \dfrac{4}{1,5} = 2,67 \ \Omega$

(c) $V_2 = IR = 1,5 \times 3 = 4,5 \ V$

(d) $E = V_1 + V_2 = 4 + 4,5 = 8,5 \ V$

$\therefore$ E of each cell $= \dfrac{8,5}{2} = 4,25 \ V$

(e) $Q = It = 1,5 \times (2 \times 60) = 180 \ C$

(f) $W = VQ = VIt = 4,5 \times 1,5 \times (2 \times 60) = 810 \ C$

76

Example 3:

A 12 V battery is connected in a circuit as shown on the diagram. Calculate the:

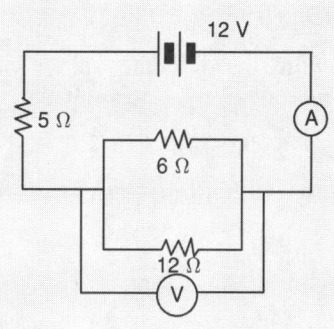

(a) total resistance of the circuit.
(b) reading on the ammeter.
(c) reading on the voltmeter.
(d) current flowing through the 6 Ω resistor.

Solution:

(a) $\dfrac{1}{R_{\parallel}} = \dfrac{1}{6} + \dfrac{1}{12} = \dfrac{2+1}{12}$ ∴ $R_{\parallel} = 4\ \Omega$

  $R_{tot} = 5 + 4 = 9\ \Omega$

(b) $I = \dfrac{V}{R} = \dfrac{12}{9} = 1,33\ A$

(c) $V_{\parallel} = IR_{\parallel} = 1,33 \times 4 = 5,33\ V$

(d) $I = \dfrac{V}{R} = \dfrac{5,33}{6} = 0,89\ A$

Example 4:
An electric circuit as shown in the diagram, is connected.

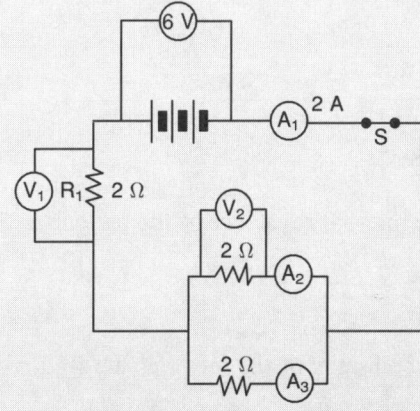

(a) Is the voltmeter reading of 6 V the emf or the potential difference of the circuit? Explain.
(b) What is the amount of charge flowing through the circuit per second?
(c) Calculate the readings on:
   (i) A_2
   (ii) A_3
   (iii) V_1
   (iv) V_2

(d) How much work is needed to let a charge of 3 C flow through resistor R_1?

Solution

(a) Potential difference of the circuit because the switch S is closed ⇒ a current flows (the voltmeter measures the emf when no current is flowing, i.e. when the circuit is open)

(b) $Q = It = 2 \times 1 = 2$ C

(c) (i) Current divides equally between the two 2 Ω-resistors: $I = \dfrac{2}{2} = 1$ A at A_1 and A_2.
(ii) see (i)
(iii) $V_1 = IR_1 = 2 \times 2 = 4$ V
(iv) $V_2 = I_2 \times R = 1 \times 2 = 2$ V

(d) $W = V_1 Q = 4 \times 3 = 12$ J

Example 5:

In the following circuit each cell has an internal resistance of 0,5 Ω and an emf of 6 V.

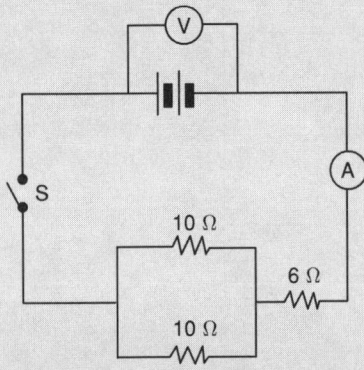

(a) Explain what is (i) internal resistance, and
(ii) lost volts.
(b) What is the reading on V when switch S is open?
S is now closed.
(c) Calculate the total resistance of the circuit (the internal resistance of the battery *included*).
(d) Calculate the reading on the ammeter.

Solution

(a) (i) Resistance exerted by the cell due to the collisions of the moving charges inside the cell.
(ii) the amount of energy per coulomb charge transferred to the cell as a result of the internal resistance of the cell

(b) 12 V

(c) $\dfrac{1}{R_\parallel} = \dfrac{1}{10} + \dfrac{1}{10} = \dfrac{1}{5}$ ∴ $R_\parallel = 5\,\Omega$

$R_{tot} = 6 + 5 = 11\,\Omega$

(d) $I = \dfrac{V}{R} = \dfrac{12}{11 + 1} = 1$ A

QUESTIONS

Section A

Various possibilities are suggested as answers to the following questions. Indicate the correct answer.

1. The rate of flow of 1 C of charge in a circuit, is a measure of the . . .
   - A resistance.
   - B potential difference.
   - C electrical power. ✓
   - D electric current.

2. Which SI unit measures the rate of flow of electric charge?
   - A Watt
   - B Coulomb ✓
   - C Volt
   - D Ampère

3. The unit of electrical resistance is . . .
   - A volt. ✓
   - B joule.
   - C ohm.
   - D watt.

4. One volt is the same as one . . .
   - A joule/ampère.
   - B joule/coulomb. ✓
   - C coulomb/joule.
   - D ampère/coulomb.

5. Choose the correct units for the following electrical quantities:

	Current	Charge	Potential difference	Resistance
A	ampère ✓	coulomb ✓	ohm	ohm ✓
B	volt.	coulomb ✓	ohm	ampère
C	coulomb.	volt	volt ✓	volt
D	ampère. ✓	coulomb ✓	ampère	ohm ✓

6. A voltmeter is connected in turn across different components in a circuit. The reading . . .
   - A always indicates the potential difference of the cell. ✓
   - B is always the same.
   - C is an indication of the amount of energy transferred by the moving charges.
   - D is not influenced by the resistance of the specific components.

7. A factor which does **not** affect the resistance of a resistor, is the . . .
   - A cross-sectional area.
   - B length of the conductor.
   - C type of material from which it is made.
   - D potential difference across the conductor. ✓

8. The magnitude of the electric current in a circuit will definitely increase if . . .
   - A more cells are connected in parallel.
   - B the resistors are connected in parallel.
   - C the cells are connected in parallel and the resistors in series.
   - D both the cells and resistors are connected in parallel. ✓

9. Which one of the following circuits can be used to calculate the resistance of the resistor?

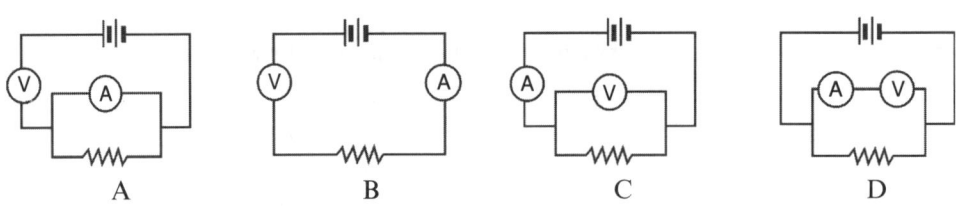

A   B   C   D

10. The following diagrams represent different connections of 2 V cells. Which connection will produce a potential difference of 4 V?

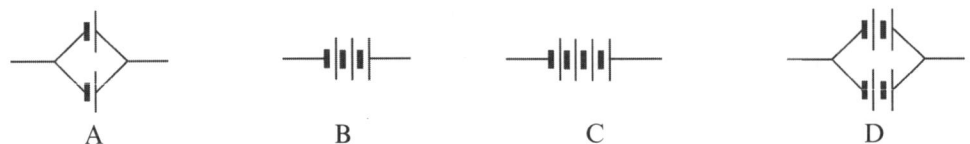

A   B   C   D

11. If a main current of 3 A is divided by the accompanying circuit, the current flowing through the
4 Ω resistor is . . .

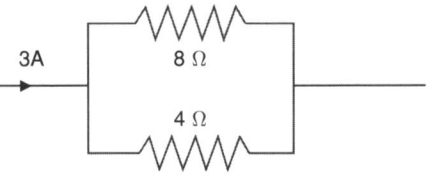

A 3 A.
B 2 A.
C 1 A.
D 1,5 A.

12. The resistance against the flow of charge inside a cell, is the . . .
    A lost volts.   C internal resistance.
    B effective resistance.   D internal volts.

13. A voltmeter which is connected across a bulb in a circuit, reads 2 V. Consider the following:
    For every one coulomb of charge flowing through the bulb, . . .

    I   2 J of heat and light is liberated
    II   the potential energy decreases with 2 J.
    III   the potential energy increases with 2 J.

    The correct statements are . . .

    A I and II.   C II and III.
    B I and III.   D I, II and III.

14. An electric toaster is connected to a 200 V source for 1 minute. The current in the toaster is
    2 A. The work done in the element of the toaster is . . .
    A 400 J.   C 24 000 J.
    B 1 200 J.   D 600 J.

15. Which one of the following connections of a number of 4 Ω-resistors will give an effective
    resistance of 4 Ω

    A   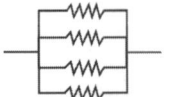   C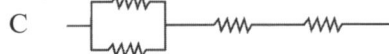

    B   D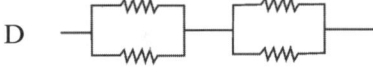

16. Three resistors, 3 Ω each, are connected to a 6 V battery. The connection which produces a current of 6 A, is . . .

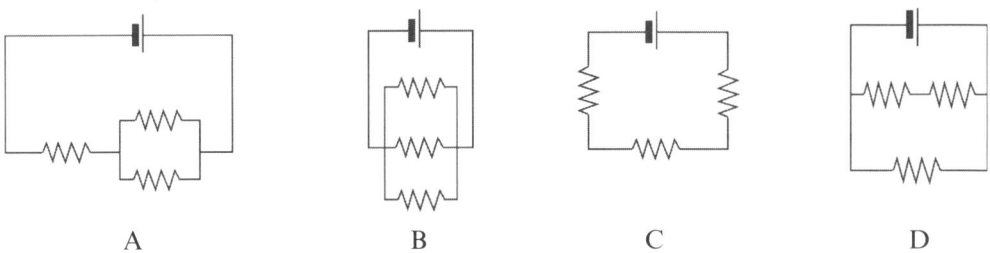

| A | B | C | D |

17. Which one of the following circuits has an effective resistance of 10 Ω?

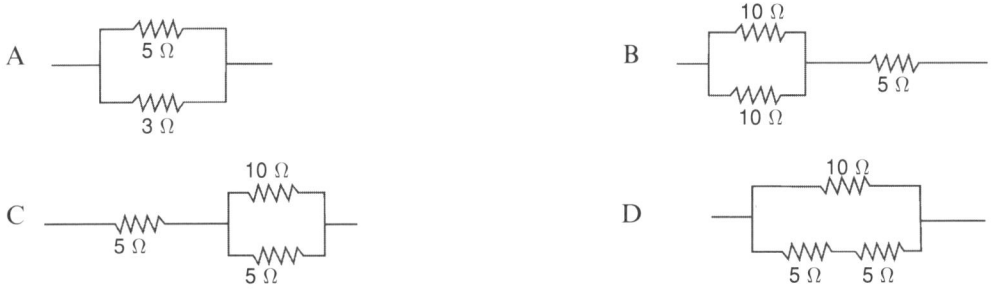

Additional for Higher Grade

18. Study the diagram. If the positions of the ammeter (A) and the voltmeter (V) are accidentally changed around, which of the following will describe the readings the best?

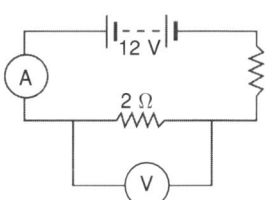

	Ammeter	Voltmeter
A	zero reading	zero reading
B	zero reading	reads 12 V
C	very large reading	zero reading
D	zero reading	very small reading

19. Bulb A is connected as shown in the diagram.
If the filament of bulb A burns out, the potential difference across the bulb will become . . .
A zero.
B infinitely large.
C the same as before.
D equal to the emf of the cell.

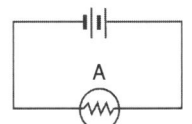

20. Four torch cells of 1,5 V each, are connected in series to form a battery of cells. This battery is connected to a resistor which has a resistance of 5 ohm. The current through the resistor is
A 0,3 A.   C 2 A.
B 1,2 A.   D 7,5 A.

21. In the accompanying circuit diagram, the 5 V battery has a very small internal resistance and X is an unknown component. The other values are shown.

Component X should be a/an ...

A ammeter.
B cell.
C voltmeter.
D resistor.

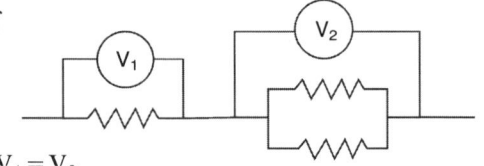

22. The diagram represents a section of an electric circuit. The resistors are identical. Which one of the following best represents the relationship between the voltmeter readings V_1 and V_2 when the current flows in the circuit?

A $V_1 = V_2 = 0$ C $V_1 = V_2$
B $V_1 = \frac{1}{2}V_2$ D $V_1 = 2V_2$

Section B

1. A charge of 100 C flows past a point in 5 s. Calculate the current in the circuit.

2.1 Define a potential difference of 1 V.
2.2 Which instrument is used to measure potential difference and in what unit is it measured?
2.3 How must this instrument be connected in a circuit to measure the potential difference?
2.4 What does it mean when a voltmeter shows a reading of 6 V?

3. Discuss why a voltmeter in a circuit is connected in parallel over a component and not in series.

4. You have two meters which are not marked. One is a voltmeter and the other an ammeter. Discuss how you will go about to determine which meter is which.

5. How does the resistance of an ammeter compare to that of a voltmeter? Explain your answer.

6. The potential difference between two points in a circuit is 6 V. How much work is needed to move a charge of 1 C between the points?

7. A torch cell is "flat". Explain what is meant by this.

8. To move 1 C from point A to point B as shown in the sketch, 2,5 J of energy is needed. Determine:

8.1 the potential difference across points A and B;

8.2 the reading on the voltmeter;

8.3 the current flowing in the circuit if the resistance of the resistor is 5 Ω.

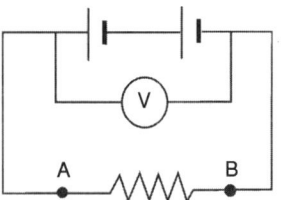

9. A cell in a circuit has a potential difference of 1,5 V across its terminals.

9.1 How much work need to be done by the cell to allow a charge of 3 C to flow through the circuit?

9.2 What will be the strength of the current if it takes 2 s to send the charge of 3 C through the circuit?

10. Four resistors of 5 Ω, 10 Ω, 20 Ω and 25 Ω respectively, are connected in series between the terminals of a 24 V battery.

10.1 Draw a diagram of the circuit.

10.2 Calculate the total resistance of the resistors in series.

10.3 Calculate the current flowing in the circuit.

10.4 Calculate the amount of charge flowing through the 10 Ω resistor in two minutes.

11. Two resistors, each with a resistance of 4 Ω, are connected in a circuit. What is the effective resistance of the circuit if they are connected in

11.1 series?

11.2 parallel?

12. Two identical bulbs L_1 and L_2 in parallel and a resistor R are connected in series to a 1,5 V cell as shown in the sketch.

Redraw the sketch and add the following components:

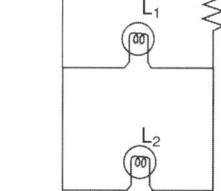

12.1 Five additional cells of 1,5 V each for a total of 6 cells with a total emf of 3 V.

12.2 An ammeter which shows the total current.

12.3 A voltmeter which shows the potential difference over both L_1 and L_2.

12.4 A switch which will increase the current when it is closed.

13. A circuit consists of four 1,5 V-cells in series, a switch, an ammeter in the main circuit and two resistors of 2 Ω each, connected in parallel. The reading on the ammeter is 6 A.

13.1 Draw a diagram of the circuit.

13.2 In your diagram, show how a voltmeter must be connected to measure the potential difference across the two resistors.

13.3 Calculate the total resistance of the resistors.

13.4 Calculate the charge flowing through both resistors in two minutes.

13.5 Calculate the reading on the voltmeter.

13.6 Calculate the work done by both resistors if a charge of 360 C flows through them.

14. A current of 1,5 A flows through the accompanying circuit when a battery with an emf of 3 V is connected in series.

Calculate

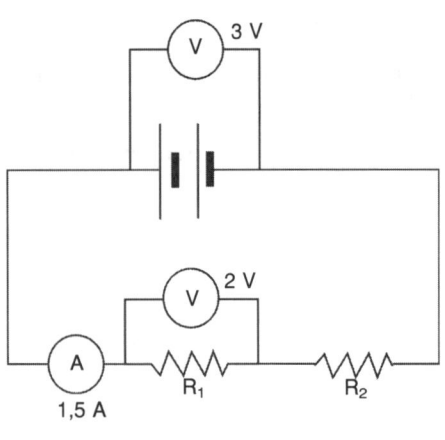

14.1 the amount of charge flowing through the ammeter in 10 seconds.

14.2 the resistance of R_1.

14.3 the resistance of R_2.

14.4 the amount of energy transferred to resistor R_2 in 5 minutes.

15. For the following circuit, calculate

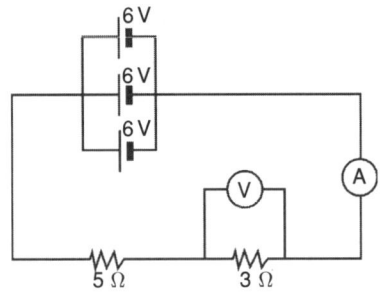

15.1 the total resistance of the circuit.

15.2 the reading on the ammeter.

15.3 the reading on the voltmeter.

15.4 the amount of charge flowing through the 5 Ω resistor in 5 minutes.

16. A learner is asked to determine the resistance of resistor X in the following circuit. He is provided with an ammeter, a voltmeter and three cells. He connects the ammeter and voltmeter in the correct places, and then take readings first for one cell, then two and then three cells connected in series.

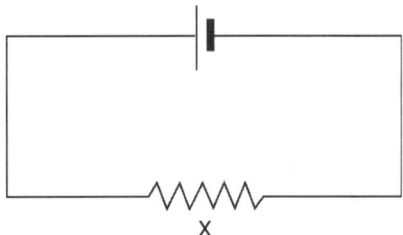

16.1 Redraw the circuit diagram and show how the ammeter and voltmeter must be connected to determine the resistance of X.

The following readings are obtained:

Number of cells	1	2	3
Current (ampère)	0,44	0,86	1,3
Potential difference (volt)	1,5	3,0	4,5

16.2 Draw a sketch graph of potential difference versus current from the above set of readings.

16.3 What can be deduced from the shape of the graph?

16.4 Calculate the magnitude of resistor X.

17. The potential difference across a resistor is 6 V and the current through it 2 A.

17.1 Calculate the resistance of the resistor.

17.2 How will the resistance change if the temperature of the resistor increases?

17.3 How will the current through the resistor change if the potential difference across the resistor is doubled without changing the temperature?

17.4 Why can this deduction be made?

Additional for Higher Grade

18. When switch S in the accompanying circuit is closed, the reading on the ammeter is 2 A, and the readings on voltmeters V_1 and V_2 are equal.

Calculate

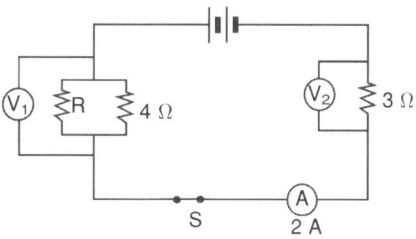

18.1 each voltmeter reading.

18.2 the potential difference accross the battery.

18.3 the resistance of resistor R.

19. In the accompanying circuit, calculate

19.1 the resistance of the parallel connection.

19.2 the reading on the ammeter.

19.3 the reading on voltmeter V_1.

19.4 the amount of heat liberated by the 3 Ω resistor in 5 minutes.

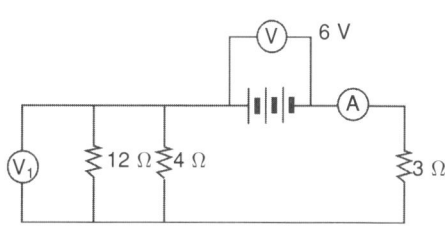

20.1 Explain what is meant by the emf of a cell. A voltmeter is connected across the terminals of a cell as shown in the sketch. S is a switch and R a resistor

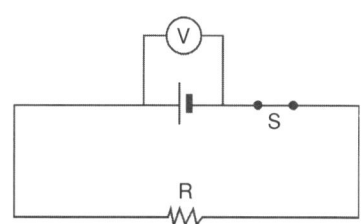

20.2 When can the reading on the voltmeter be regarded as
A the emf of the cell, and
B the potential difference?

20.3 Which of the emf or the potential difference has the greater value? Explain your answer.

21. The cell in the circuit has an emf of 10 V. The reading on the voltmeter is 5 V. The resistors are indicated on the diagram.

21.1 Calculate the reading on A_1.

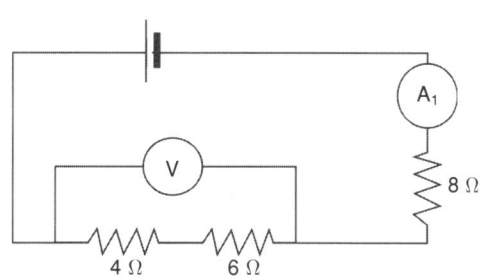

21.2 Calculate the potential difference across all the resistors.

21.3 Explain the difference between the answer in question 21.2 and the emf of the cell.

21.4 Calculate the charge flowing through the 8 Ω resistor in 5 minutes.

ANSWERS

Section A

1. D	**2.** D	**3.** C	**4.** B	**5.** D	**6.** C	**7.** D
8. B	**9.** C	**10.** D	**11.** B	**12.** C	**13.** A	**14.** C
15. D	**16.** B	**17.** B				

Additional for Higher Grade

18. B	**19.** D	**20.** B	**21.** B	**22.** D

Section B

1. $I = \dfrac{Q}{t} = \dfrac{100}{5} = 20$ A

2.1 A potential difference of 1 V exists across a conductor if 1 J of energy is transferred to the conductor when a charge of 1 C flows through it.

2.2 voltmeter; volt (V)

2.3 in parallel

2.4 6 J of energy is converted to heat or light by the conductor when a charge of 1 C flows through it.

3. A voltmeter measures the difference in energy between two points; thus it must be connected between two points (A and B)
$\Rightarrow$ therefore in parallel

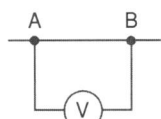

4. Connect a circuit with one cell and the meter in series. If it gives a reading, it is an ammeter; if it gives no reading, it is a voltmeter.

5. voltmeter: great resistance, no current should flow through it.
ammeter: small resistance; it is connected in series in a circuit to allow the total current to flow through it.

6. $W = VQ = 6 \times 1 = 6$ J

7. The chemical reaction in the cell which supplies the electrical energy, is completed. Thus no electrical energy can be transferred to the charges.

8. **8.1** $V = \dfrac{W}{Q} = \dfrac{2,5}{1} = 2,5 \text{ V}$

8.2 $2,5 \text{ V}$

8.3 $I = \dfrac{V}{R} = \dfrac{2,5}{5} = 0,5 \text{ A}$

9. **9.1** $W = VQ = 1,5 \times 3 = 4,5 \text{ J}$

9.2 $I = \dfrac{Q}{t} = \dfrac{3}{2} = 1,5 \text{ A}$

10.1

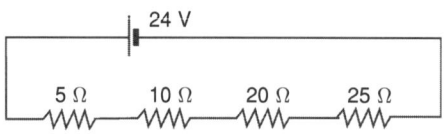

10.2 $R_{tot} = 5 + 10 + 20 + 25 = 60 \ \Omega$

10.3 $I = \dfrac{V}{R} = \dfrac{24}{60} = 0,4 \text{ A}$

10.4 $Q = It = 0,4 \times (2 \times 60) = 48 \text{ C}$

11.1 4 Ω   4 Ω $\therefore \quad R_{tot} = 4 + 4 = 8 \ \Omega$

11.2 4 Ω $\therefore \quad \dfrac{1}{R_{tot}} = \dfrac{1}{4} + \dfrac{1}{4} = \dfrac{2}{4} = \dfrac{1}{2}$

4 Ω $\therefore \quad R_{tot} = \dfrac{2}{1} = 2 \ \Omega$

12.

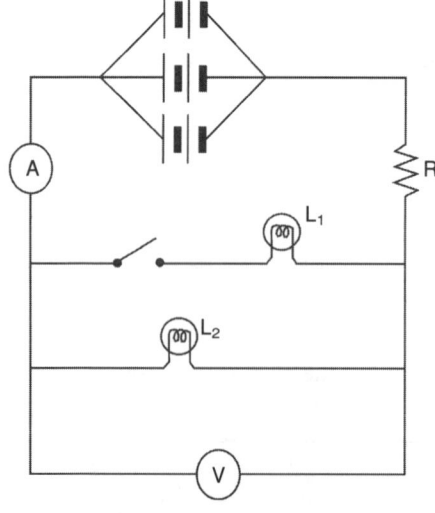

13.1

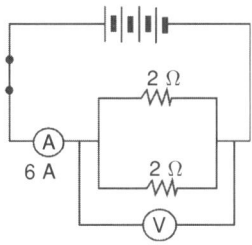

13.2 see sketch

13.3 $\dfrac{1}{R_{tot}} = \dfrac{1}{2} + \dfrac{1}{2} = \dfrac{2}{2}$ $\therefore$ $R_{tot} = 1\ \Omega$

13.4 $Q = It = 6 \times (2 \times 60) = 720\ C$

13.5 $V = IR = 6 \times 1 = 6\ V$

13.6 $W = VQ = 6 \times 360 = 2\,160\ J$

14.1 $Q = It = 1,5 \times 10 = 15\ C$

14.2 $R = \dfrac{V}{I} = \dfrac{2}{1,5} = 1,33\ \Omega$

14.3 $R_2 = \dfrac{V}{I} = \dfrac{(3-2)}{1,5} = 0,67\ \Omega$

14.4 $W = QV = (It)V = 1,5 \times (5 \times 60) \times 1 = 450\ C$

15.1 $R_{tot} = 5 + 3 = 8\ \Omega$

15.2 $I = \dfrac{V_{tot}}{R_{tot}} = \dfrac{6}{8} = 0,75\ A$

15.3 $V = IR = 0,75 \times 3 = 2,25\ V$

15.4 $Q = It = 0,75 \times (5 \times 60) = 225\ C$

16.1

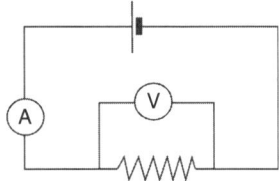

16.2

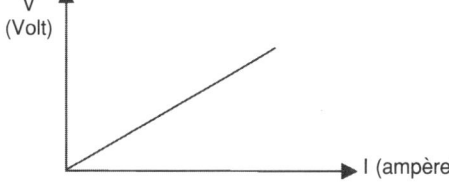

16.3 a straight line which indicates that the voltmeter readings are directly proportional to the ammeter readings

16.4 1 cell: $R = \dfrac{V}{I} = \dfrac{1,5}{0,44} = 3,41\ \Omega$

2 cells: $R = \dfrac{V}{I} = \dfrac{3}{0,86} = 3,49\ \Omega$

3 cells: $R = \dfrac{V}{I} = \dfrac{4,5}{1,3} = 3,46\ \Omega$

average value: 3,45 Ω

17.1 $R = \dfrac{V}{I} = \dfrac{6}{2} = 3\ \Omega$

17.2 resistance increases; when the temperature increases, the atoms of the conductor vibrate more vigorously causing the charges to move more difficult through the conductor.

17.3 current is doubled

17.4 V is directly proportional to I

Additional for Higher Grade

18.1 $V_2 = IR = 2 \times 3 = 6\ V$

$V_1 = V_2 = 6\ V$

18.2 $E = 6 + 6 = 12\ V$

18.3 $R_\parallel = \dfrac{V_1}{I} = \dfrac{6}{2} = 3\ \Omega$

$\therefore \dfrac{1}{R_{tot}} = \dfrac{1}{R} + \dfrac{1}{4}$

$\therefore \dfrac{1}{3} = \dfrac{1}{R} + \dfrac{1}{4}$

$\therefore \dfrac{1}{R} = \dfrac{1}{3} - \dfrac{1}{4} \quad \therefore \ R = 12\ \Omega$

19.1 $\dfrac{1}{R_\parallel} = \dfrac{1}{12} + \dfrac{1}{4} = \dfrac{4}{12} \quad \therefore \ R_\parallel = \dfrac{12}{4} = 3\ \Omega$

19.2 $R_{tot} = 3 + 3 = 6\ \Omega$

$I = \dfrac{V_{tot}}{R_{tot}} = \dfrac{6}{6} = 1\ A$

19.3 $V_1 = IR_\parallel = 1 \times 3 = 3\ V$

19.4 $W = VQ = V(It) = (6 - 3) \times 1 \times (5 \times 60) = 900\ J$

20.1 Maximum amount of electrical energy which a cell can produce per coulomb charge.

20.2 A: when switch S is open (and no current is flowing).
B: when switch S is closed and a current does flow.

20.3 emf > potential difference

- When the switch is closed and a current starts flowing,
- energy is transferred to the cell (the "lost volt")
- due to the internal resistance of the cell.
- the reading on the voltmeter decreases
- because it measures the energy transferred to the external circuit.

21.1 $I = \dfrac{V}{R} = \dfrac{5}{10} = 0,5$ A $\qquad (R = 4 + 6 = 10 \ \Omega)$

21.2 $\qquad V_{8 \ \Omega} = IR = 0,5 \times 8 = 4$ V

$\qquad \therefore \quad V_{system} = 5 + 4 = 9$ V

21.3
- There is a difference of 1 V between the emf of the cell and the answer in 21.2.
- It is a result of the internal resistance of the cell.
- Energy is lost in the cell in the form of heat when the current flows through it.

21.4 $Q = It = 0,5 \times (5 \times 60) = 150$ C.

6 The Effects of an Electric Current

1. Heating effect of an electric current

When a **potential difference** is applied across the ends of a conductor, **energy is given out** in the form of **heat**.

1.1 The factors that determine the heat produced by a resistor

• **The electric current**
The greater the current the greater is the heat produced.

• **The resistance of the conductor**
The greater the resistance, the greater is the heat produced.

• **The time that the current flows**
The longer the time that the current flows, the greater is the heat produced.

1.2 Energy conversion during electric heating

Electrons move through the conductor when current flows. The electrons receive **potential energy** from the electric source of energy (cell, dynamo, etc.). As they move away from the negative terminal, the **potential energy** is **converted** into **kinetic energy**. The electrons collide with other particles in the conductor. The kinetic energy is transferred to these particles. The **average kinetic energy** of these particles are **raised**. This causes an **increase in temperature**.

1.3 Examples of the use of the heating effect of an electric current in everyday life

All heating appliances e.g. stoves, kettles, toasters, heaters, etc. convert electrical energy into heat energy. Heating elements are usually made of nichrome which has a high resistance and a high melting point.

Electric light bulbs are also an application of the heating effect. The filament of the bulb needs to be white hot in order to give out white light. The filament of most light bulbs is made of a coiled **tungsten** wire.

1.4 The effect of temperature on the resistance of a resistor

Experiment 1

To investigate the effect of temperature on the resistance of a conductor.
• Set up a circuit as shown in fig. 1.
• AB is a coil of about 30 cm of iron wire.
• Take the readings on both the ammeter and voltmeter.

- Heat AB strongly with a bunsen burner.
- Take the new readings on the ammeter and voltmeter.

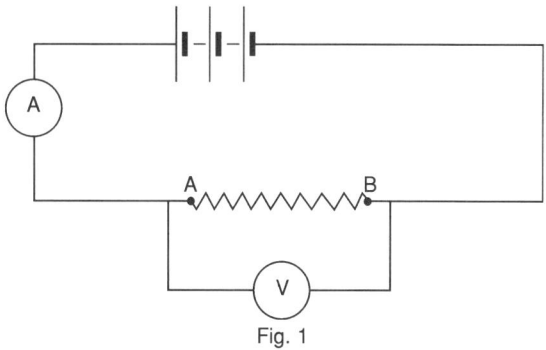

Fig. 1

Typical results

Temperature	Ammeter reading	Voltmeter reading	Resistance ($R = \dfrac{V}{I}$)
Room temperature	1A	6 V	R = 6 Ω
Hot	0,25 A	6,5 V	R = 26 Ω

Deduction
- $\dfrac{V}{I}$ = R is greater for the higher temperature. As the temperature of a conductor increases, so its resistance increases.
- The current through a conductor is determined by its temperature. As the temperature of a resistor increases, so the resistance increases. (This is only noticeable at very high temperatures.)

2. The magnetic effect of an electric current

In 1819 Hans Oersted discovered that a current in a wire causes a nearby free moving magnet to deflect.

A **magnetic field** is represented by a collection of lines called **magnetic field lines**. The strength of the magnetic field is determined by the density of the lines.

> The direction of a magnetic field is defined as the direction in which the N pole of a compass needle points when it is placed in the field.

2.1 The magnetic field of a straight current-carrying conductor

Experiment 2

To investigate the magnetic field around a straight current-carrying conductor.
- Clamp a square piece of cardboard in position as shown. The copper wire should pass vertically through the cardboard.
- Sprinkle a thin layer of iron filings on the cardboard. Tap the board gently (fig. 2(a)).
- Use small compass needles to determine the direction of the magnetic field (fig. 2(b)).

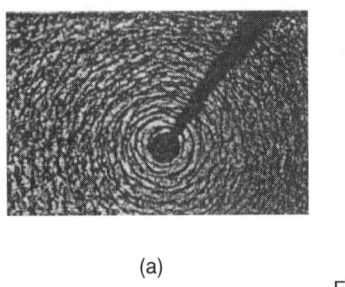

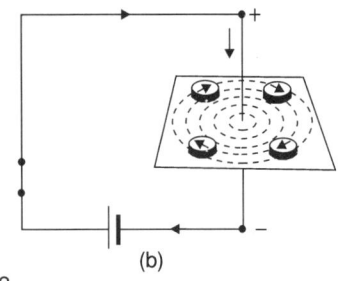

(a)	(b)

Fig. 2

Typical results and deductions

- The iron filings arrange themselves in a series of concentric circles around the wire and in a plane at right angles to it as shown in fig. 2(a). The pattern is most pronounced closest to the wire, indicating that the **magnetic field is strongest closest to the wire**.
- The direction of the field lines can be determined by using the **right hand wire rule**.

Hold the wire in your right hand, with your thumb pointing in the direction of the conventional current. The curved fingers indicate the direction of the magnetic field (fig. 3).

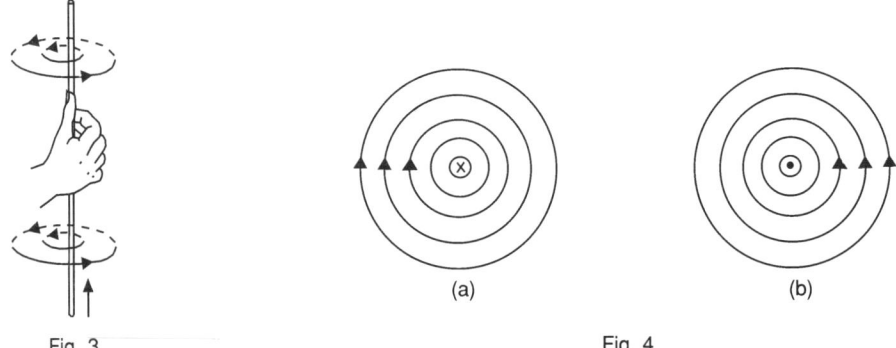

Fig. 3   Fig. 4

- The magnetic field pattern can be represented by field lines drawn in one plane. The symbol ⊙ indicates an approaching current and the symbol ⊗ a current moving away from you into the page (fig. 4).

2.2 The magnetic field due to a circular conductor

Experiment 3

To investigate the magnetic field around a circular current-carrying conductor.
- Set up a circuit as shown in fig. 5.
- Sprinkle a thin layer of iron filings on the cardboard and tap the board gently (fig. 6).
- Use small compass needles to determine the direction of the magnetic field.

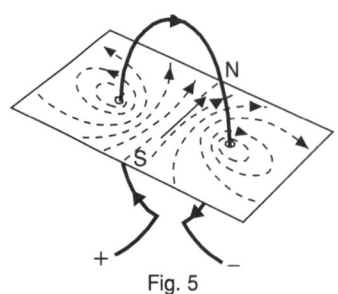

Fig. 5

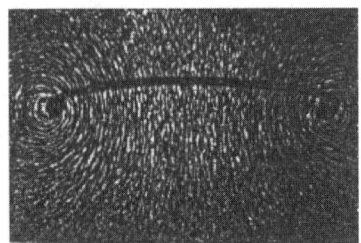

Fig. 6

Typical result and deductions

- The magnetic field lines closest to the wire are **circular**. At the centre of the coil the field lines are **straight and perpendicular** to the plane of the coil.
- The direction of the field can be determined by applying the **right-hand rule**:

  Grasp any section of the coil with your right hand so that your extended thumb points in the direction of the current. Your curved fingers indicate the direction of the magnetic field.

- The magnetic field pattern can be represented by lines drawn in one plane (fig. 7).

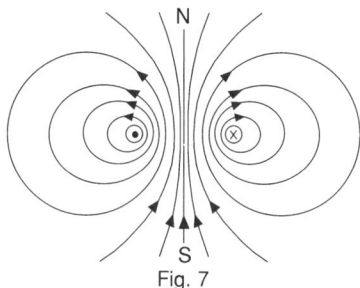

Fig. 7

2.3 The magnetic field due to a solenoid

Experiment 4

To investigate the magnetic field around a current carrying solenoid.
- Set up a circuit as shown in fig. 8.
- Sprinkle a thin layer of iron filings on the cardboard and tap the board gently (fig. 9).
- Use small compass needles to determine the direction of the magnetic field.

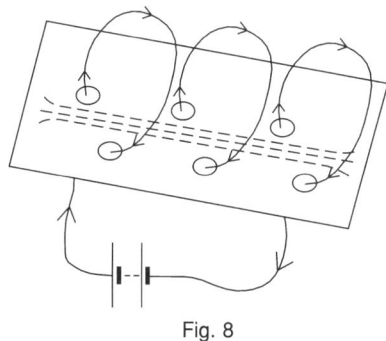

Fig. 8

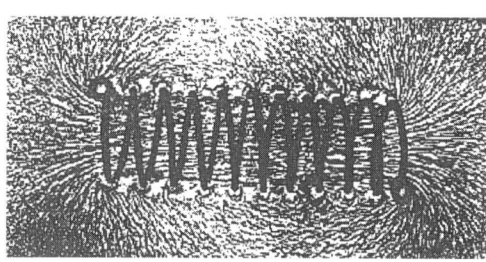

Fig. 9

Typical results and deductions

- The magnetic field around a solenoid looks almost similar to that around a bar magnet.
- The **direction of the field outside the solenoid is from N to S**. The field **inside the solenoid** is from S to N. (The magnetic field lines must form closed loops.)
- The polarity of the current-carrying solenoid can be determined by the **right-hand solenoid rule**.

If the solenoid is held in the right hand with the curved fingers pointing in the direction of the conventional current, the extended thumb points in the direction of the N pole of the solenoid (fig. 10).

• The pattern of the field can be drawn in one plane (fig. 11).

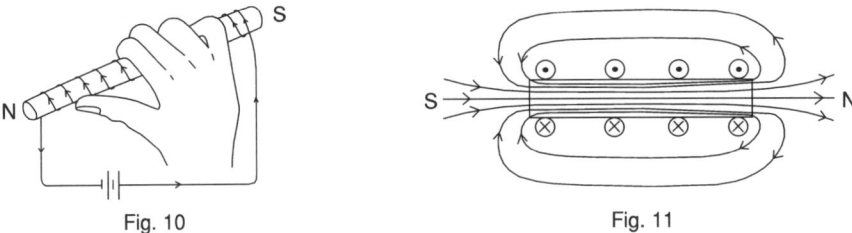

Fig. 10   Fig. 11

Important! The iron filings and all the diagrams show the presence of the magnetic fields in one plane only, namely the plane of the paper. The field exists in three dimensions along the entire length of the conductor.

3. Applications of the magnetic effect

3.1 The electromagnet

A solenoid in which a piece of iron is placed is called an **electromagnet**. It has magnetic properties only as long as a current flows in the solenoid.

Magnetic materials

• **Soft magnetic materials** e.g. **iron** is easily magnetised, but loses its magnetism when the current is switched off.
• **Hard magnetic materials** e.g. **steel** is not easily magnetised, but it retains its magnetic properties after the current has been switched off.

The strength of an electromagnet can be increased by

• increasing the current
• increasing the number of windings.

The functioning of an electromagnet depends on the type of magnetic material used in the core of the solenoid.

Electromagnets are used as lifting magnets, in electric bells, magnetic relays, the telephone, the magnetic circuit breaker, the microphone, etc.

3.2 The motor effect

A current-carrying conductor experiences a force when placed in a magnetic field.

Experiment 5

To investigate the influence of a magnetic field on a current-carrying conductor.
• Set up the apparatus as shown in fig. 12.
• Close the switch and note the movement of the free-hanging wire.
• Change the magnetic poles around and close the switch again.
• Change the terminals of the battery and close the switch again.

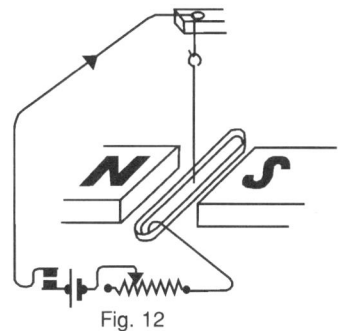

Fig. 12

Results and deductions

- The free-hanging copper conductor experiences no force when the current does not flow.
- When the **current flows**, the interaction between the magnetic field of the conductor and the magnetic field of the bar magnets causes **a force** to be exerted on the conductor and the conductor moves to one side.
- When the magnetic poles are changed, the force is reversed and the conductor swings in the opposite direction.
- The same effect is produced when the terminals are reversed so that the current flows upwards in the conductor.
- The **magnetic field**, the **electric current** and the direction of the **force** are **at right angles to one another**.
- If the conductor should be placed parallel to the field, no force is experienced. **Maximum force** is experienced when the **conductor is at right angles to the field.** A weaker force is produced for any other position.

Fleming's Left hand motor rule

Place the **forefinger, middle finger** and **thumb** of the left hand at right angles. If the forefinger points in the direction of the **field** and the middle finger in the direction of the **conventional current**, the **thumb** will point in the direction of the **thrust** of the force.

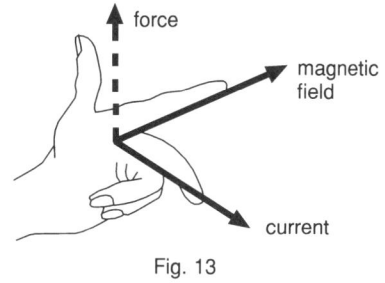

Fig. 13

This phenomenon, called the **motor effect**, is most commonly applied in electrical measuring instruments (ammeters, voltmeters, etc.) and in electric motors.

The motor effect is a very important aspect of Grade 10 work. The term "motor" refers to "motion". This was a very important breakthrough because the effect explains how **electrical energy** is converted into **energy of motion**.

3.3 The direct current electric motor

The motor effect is applied in the functioning of the electric motor. The diagram shows the basic principle of the motor effect and the left-hand motor rule. The rectangular loop rotates through 90° because part **ab** of the coil experiences a vertically downward force and part **cd** experiences a vertically upward force (Fig. 14).

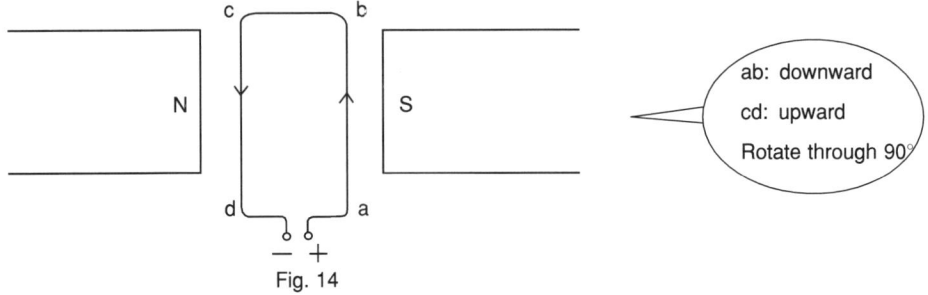

Fig. 14

In an electric motor **continuous rotation** is necessary. On reaching the vertical position, the **momentum** of the coil carries it past that position and the **current changes direction**. This is made possible by the **split ring commutator**. Electrical contact is made with the metal split ring commutator by means of fixed carbon **brushes**. Each brush is in contact with half of the commutator during one half turn, and with the other half during the second half turn.

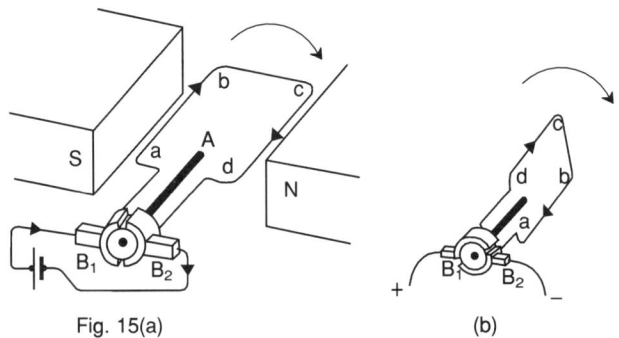

Fig. 15(a)   (b)

Consider ab:

• ab experiences an **upward force** when in the horizontal position when current flows from a to b. Fig. 15(a).
• When the loop is vertical, the **brushes bridge the gaps** in the commutator, causing the current to pass from one brush to the other.
• Very little current passes through the loop.
• The **momentum** of the loop carries it past the vertical position.
• The two commutator halves change contact from one brush to the other.
• Side ab is now on the right hand side and current flows from b to a. Fig. 15(b).
• ab is forced **downward** to the vertical position again.
• This continues as long as the current is flowing through the coil.

The following factors increase the turning force exerted on the coil of an electric motor:

• An increase in the current
• An increase in the number of windings
• An increase in the strength of the magnetic field
• An increase in the surface area of the coil
• The coil wound round soft iron, which concentrates the magnetic field.

The resultant magnetic field pattern in a simple electric motor

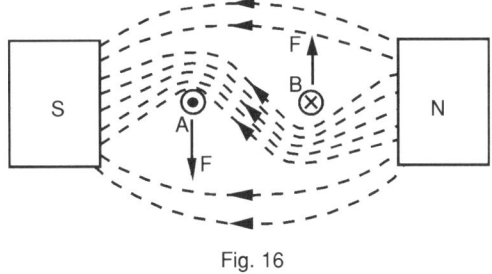

Fig. 16

The conductors are forced into the region of the weaker magnetic field.

QUESTIONS

Section A

A. *Various possibilities are suggested as answers to the following questions. Indicate the correct answer.*

1. The direction of the magnetic field at point R near the current carrying conductor is . . .

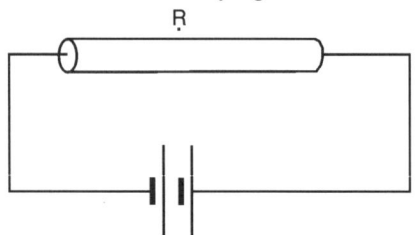

A   perpendicular to the conductor.
B   to the left.
C   vertically out of the book.
D   vertically into the book.

2. The sketch shows a magnetic field. A **negative charge** moves at point P in the direction perpendicularly into the book. The charge will experience a force in the direction of . . .

A   A.
B   B.
C   C.
D   D.

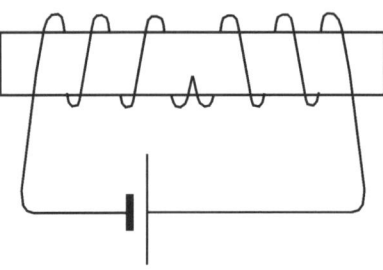

3. Electrons move in a wire. In their vicinity there will be . . .
A   an electric field only.   C   a magnetic field only.
B   both a magnetic and an electric field.   D   either a magnetic or an electric field.

4. Insulated copper wire is wrapped around a piece of soft iron wire. The sequence of the magnetic poles which are formed in the iron will be from left to right . . .

A   SNS.
B   NSS.
C   SNN.
D   NSN.

5. A resistor AB is connected in a circuit and gradually heated. If the temperature is increased in this way, . . .

A   the potential difference across the resistor will increase.
B   the current along the wire will decrease.
C   the resistance of the wire will decrease.
D   the potential difference across the wire as well as the current will increase.

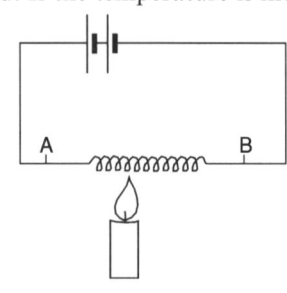

6. The diagram that represents the magnetic field of a current-carrying conductor is . . .

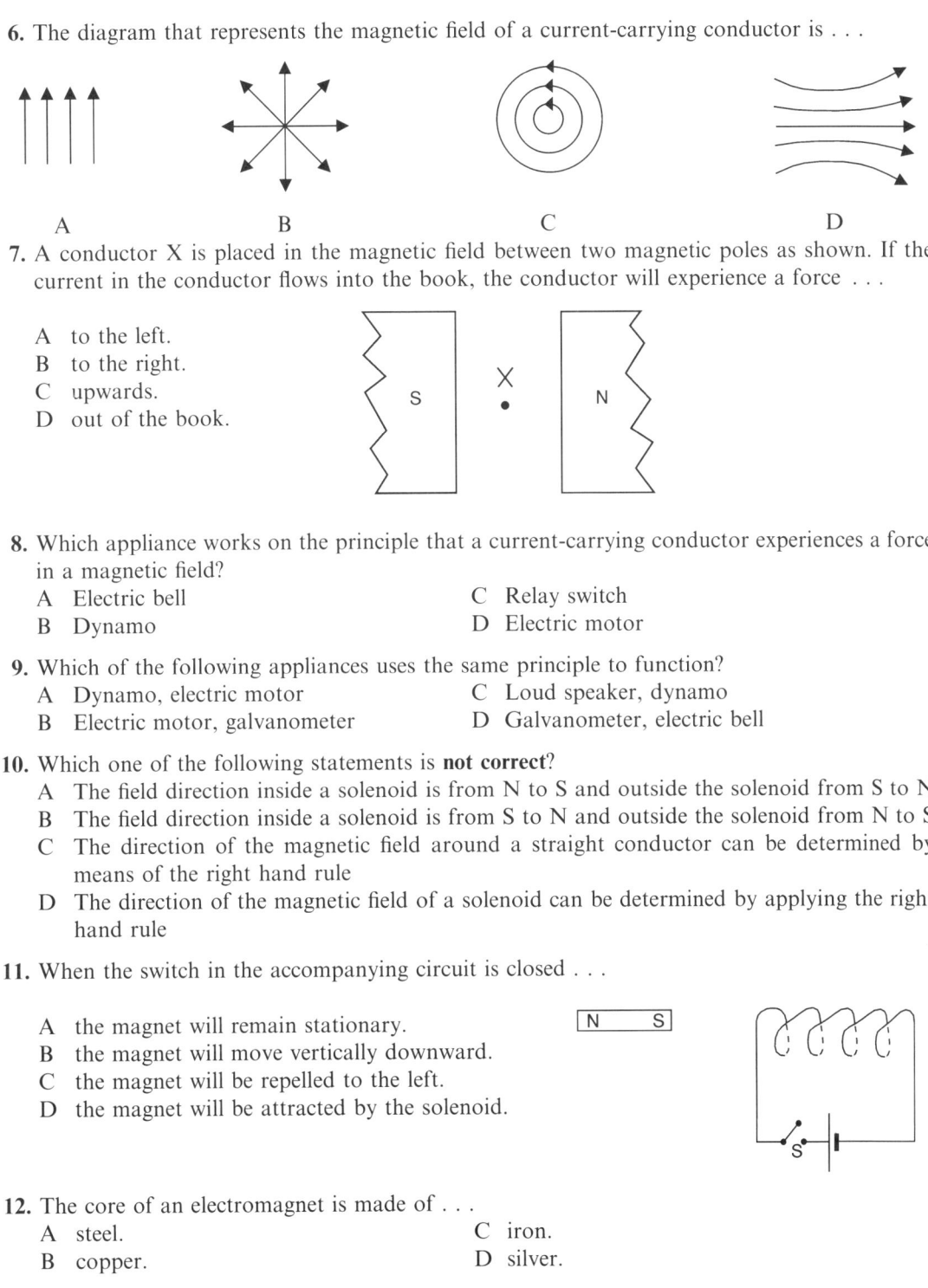

    A   B   C   D

7. A conductor X is placed in the magnetic field between two magnetic poles as shown. If the current in the conductor flows into the book, the conductor will experience a force . . .

A to the left.
B to the right.
C upwards.
D out of the book.

8. Which appliance works on the principle that a current-carrying conductor experiences a force in a magnetic field?
A Electric bell   C Relay switch
B Dynamo   D Electric motor

9. Which of the following appliances uses the same principle to function?
A Dynamo, electric motor   C Loud speaker, dynamo
B Electric motor, galvanometer   D Galvanometer, electric bell

10. Which one of the following statements is **not correct**?
A The field direction inside a solenoid is from N to S and outside the solenoid from S to N
B The field direction inside a solenoid is from S to N and outside the solenoid from N to S
C The direction of the magnetic field around a straight conductor can be determined by means of the right hand rule
D The direction of the magnetic field of a solenoid can be determined by applying the right hand rule

11. When the switch in the accompanying circuit is closed . . .

A the magnet will remain stationary.
B the magnet will move vertically downward.
C the magnet will be repelled to the left.
D the magnet will be attracted by the solenoid.

12. The core of an electromagnet is made of . . .
A steel.   C iron.
B copper.   D silver.

13. When the temperature of a conductor increases as a result of an electric current passing through it, to which effect is the following energy conversion due?
    A Chemical energy that is converted to heat energy in the wire.
    B The kinetic energy of the accelerated electrons is transferred to the particles of the conductor during collisions.
    C Electric potential energy of the moving electrons in the conductor is transferred to the particles in the wire.
    D Electric potential energy that is converted into kinetic energy in the cell.

14. If a current-carrying solenoid PQ, made of nichrome wire, is heated . . .

    A the readings on both V and A increase.
    B the reading on V decreases and the reading on A increases.
    C the readings on both V and A decreases.
    D the reading on V remains the same and the reading on A decreases.

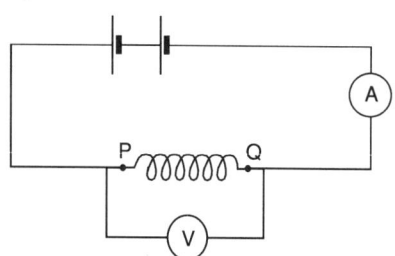

Section B

1.1 Describe the magnetic field which develops around a straight conductor, when a current flows in it.

1.2 Which rule can be used to predict the direction of this magnetic field?

1.3 State this rule.

2. Draw sketches to illustrate the shape and direction of each of the following magnetic fields:

2.1 The magnetic field around a straight conductor.

2.2 The magnetic field of a circular conductor.

2.3 The magnetic field of a solenoid.

3.1 What is an electromagnet?

3.2 Describe in short how you would make an electromagnet.

3.3 Draw a labelled sketch of an electromagnet. Indicate the current direction and the magnetic poles clearly.

3.4 Name **three** appliances in which electromagnets are used.

3.5 Name **three** ways to increase the strength of an electromagnet.

3.6 Why is the core of an electromagnet not made of steel?

3.7 What is the advantage of an electromagnet when compared to a permanent magnet?

4.1 What happens to a current-carrying conductor when placed in a magnetic field?

4.2 Name **two** instruments and/or appliances of which the functioning is based on the phenomenon mentioned in 4.1.

4.3 When considering the magnetic properties of metals, we speak of their **hardness** and **softness**. What does this mean?

5. A current-carrying conductor PQ is placed between the poles of a magnet as shown.

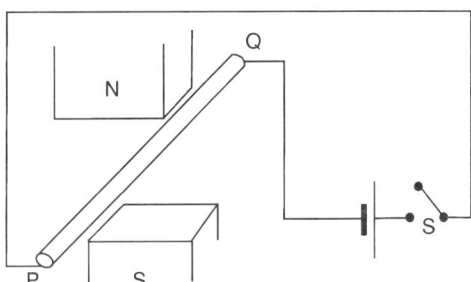

5.1 To which side will PQ swing when the switch is closed?

5.2 Name **three** ways to increase the force experienced by PQ.

5.3 Name **two** ways by which PQ can be moved to the opposite side.

5.4 To which side will the conductor move if the magnetic poles are changed around?

5.5 Which rule can be used to forecast the direction in which the conductor will move?

5.6 State this rule.

6. Two steel rods are placed inside a solenoid. An electric current is passed through the solenoid.

6.1 What will happen to the two rods? Give a reason for your answer.

6.2 Is steel a hard or a soft magnetic material?

7.

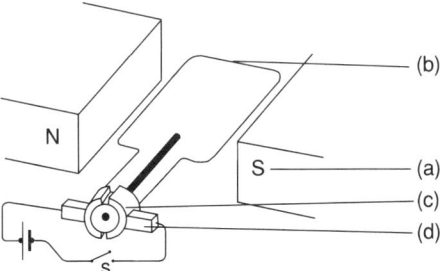

7.1 What is represented by the sketch?

7.2.1 Write down labels for the parts marked (a) to (d).

7.2.2 State the function of each of the components marked (a) to (d).

7.3   When the switch is closed, will the loop turn clockwise or anti-clockwise?

7.4   Why does the loop keep on turning after it has reached the vertical position?

Additional for Higher Grade

8.1   Draw a simple sketch to show the principle on which a moving coil galvanometer works.

8.2   Which effect is applied in the moving coil galvanometer?

8.3   Which rule can be used to predict the movement of the coil?

9.1   Sketch the resultant magnetic field of a direct current electric motor.

9.2   How could one ensure greater power and smoother running in an electric motor?

ANSWERS

Section A

1. D	**2.** A	**3.** B	**4.** A	**5.** B	**6.** C	**7.** C
8. D	**9.** B	**10.** A	**11.** D	**12.** C	**13.** B	**14.** D

Section B

1.1 • Concentric circles around the conductor in a plane at right angles to it.

    • Three dimensional along the complete length of the conductor.

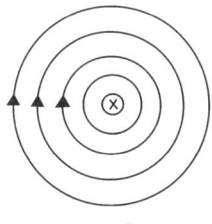 or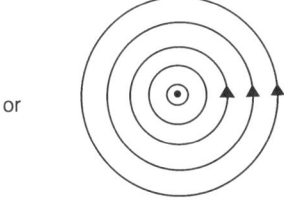

Current flows into the book   Current flows out of the book

1.2 Right-hand rule

1.3 • Hold the conductor in the right hand so that the thumb indicates the direction of the current.

    • The curved fingers indicate the direction of the magnetic field lines.

2.1

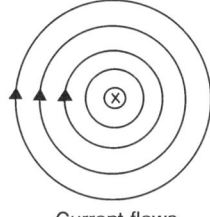 or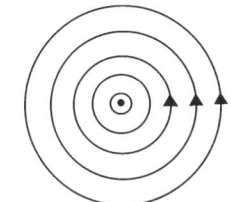

Current flows into the book   Current flows out of the book

2.2

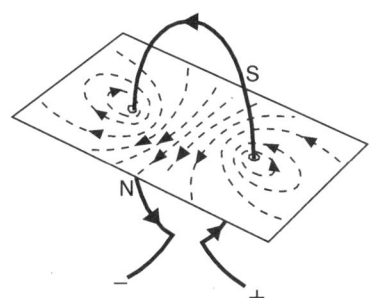

2.3

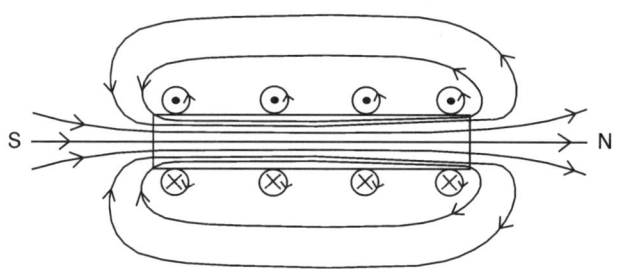

3.1 An electromagnet is a solenoid in which a piece of iron is placed. It has magnetic properties only as long as the current flows in it.

3.2 A number of windings of insulated copper wire around a piece of soft iron.

3.3

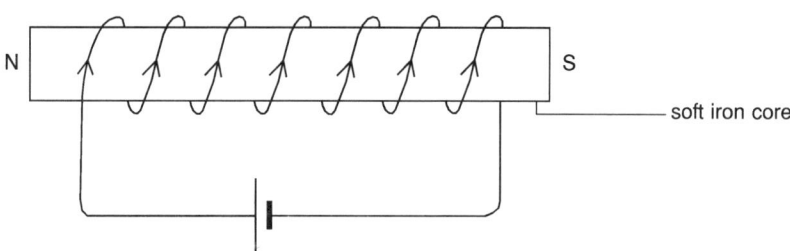

soft iron core

3.4 electric bell; loudspeaker, relay switch, etc.

3.5 • use a stronger current.

   • increase the number of windings on the core.

   • soft iron core in the solenoid.

3.6 • It does not lose its magnetism.

   • It is more difficult to magnetise because it is a hard magnetic material.

3.7 • The iron core is easily magnetised.
   • It loses all its magnetism when the current is switched off.
   • This makes it a useful component of lifting magnets, electric bells, etc. where permanent magnets would be of no use.

4.1 It experiences a force.

4.2 Electric motor; galvanometer.

4.3 **Hard magnetic materials** e.g. steel is not easily magnetised but it retains magnetic properties after the current has been switched off.

   Soft magnetic materials e.g. iron is easily magnetised but loses all its magnetism when the current is switched off.

5.1 To the left

5.2 • stronger magnet

• stronger current

• lengthen the conductor in the magnetic field

5.3 • change the magnetic poles around

• change the direction of the current around.

5.4 To the right

5.5 Fleming's Left hand motor rule.

5.6 • Put the forefinger, middle finger and thumb of the left hand at right angles.

• If the forefinger points in the direction of the field, the middle finger in the direction of the conventional current, the thumb will point in the direction of the thrust of the force.

6.1 The two rods will repel each other.

• Both rods will become magnetised when the current passes through the solenoid

• Their N poles as well as their S poles will repel each other.

6.2 Hard magnetic material.

7.1 An electric motor

7.2.1 (a) field magnets (either permanent or electromagnetic)

(b) rectangular coil made of copper wire

(c) commutator (a copper split ring)

(d) brushes (made of carbon)

7.2.2 (a) the magnets provide a permanent magnetic field

(b) the current flows through the coil which is free to rotate

(c) it reverses the current in the coil after every half turn

(d) it ensures electrical contact between the battery and the split ring commutator.

7.3 clockwise

7.4 the direction of the current changes because the brushes now make contact with the other half rings of the split ring commutator.

8.1 • The galvanometer functions on the principle that a current carrying conductor experiences a force in a magnetic field.

• The interaction between the magnetic field of the permanent magnets and the magnetic field of the current-carrying coil of the galvanometer causes the coil to move.

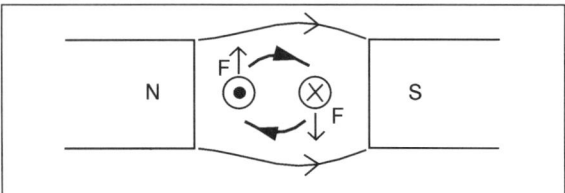

8.2 The motor-effect

8.3 Fleming's Left hand motor rule

9.1

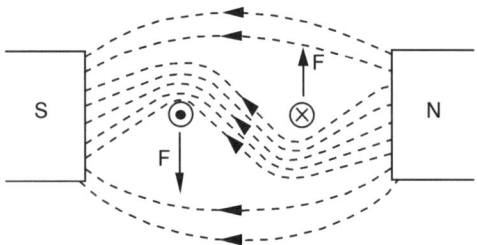

9.2 Use several coils set at different angles around an iron core.

7 *Electromagnetic Induction*

Electromagnetic induction is the phenomenon whereby a current is induced simply by moving a conductor in a magnetic field.

1. Induction of an electric current

Experiment 1:

To investigate what happens when a conductor is moved relative to a magnetic field.

- Set up a circuit as shown by using a solenoid with 150 windings and a sensitive galvanometer that has a centre zero.

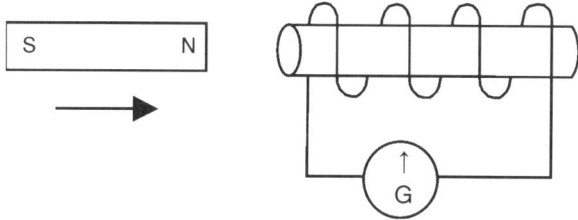

- Carry out the following investigations and observe what happens:
  - Push the N pole of a bar magnet into the coil (solenoid).
  - Hold the magnet stationary in the coil.
  - Pull the N pole out of the coil.
  - Push the S pole into the coil; turn the magnet around inside the coil; pull the magnet out.

- Keep the magnet stationary and move the coil up and down.

- **Carry out the following investigations to determine the factors that influence the strength of the current produced.**
  - Vary the speed with which you move the magnet.
  - Vary the strength of the magnet by using 2 and 3 magnets side by side, like poles touching.
  - Vary the number of turns by using solenoids with 240 and 360 windings respectively.

Results and deductions

- The galvanometer indicates that a **current is induced** in the coil only when there is **motion of the coil relative to the magnet**. When the magnet is held stationary, **no current** is produced.

- The direction of the current depends on which pole of the magnet is being used and whether it is being removed or inserted.

- **Relative motion** may be produced either by moving the magnet in the solenoid, or by moving the solenoid over the magnet.

- **The factors that influence the magnitude of the induced current are**
  - (a) **the strength of the magnetic field** – the stronger the magnet, the greater the induced current.
  - (b) **the number of turns on the coil** – the greater the number of turns, the greater the induced current.
  - (c) **the speed at which the magnet and solenoid are moved relative to each other** – the faster the movement, the greater the induced current.

> It is important to note that electromagnetic induction occurs even if the ends of the solenoid are not connected. An **induced potential difference** is produced which **causes a current** to flow when the ends of the circuit are connected.

Faraday's Law of electromagnetic induction:

> Whenever there is a change in the magnetic field linked with the conductor, a potential difference is induced. The magnitude of this potential difference is proportional to the rate of change in the magnetic flux linkage with the conductor.

2. The direction of the induced current; Lenz's Law

Lenz's Law:

> The induced current flows in such a direction that the magnetic field it produces opposes the inducing action.

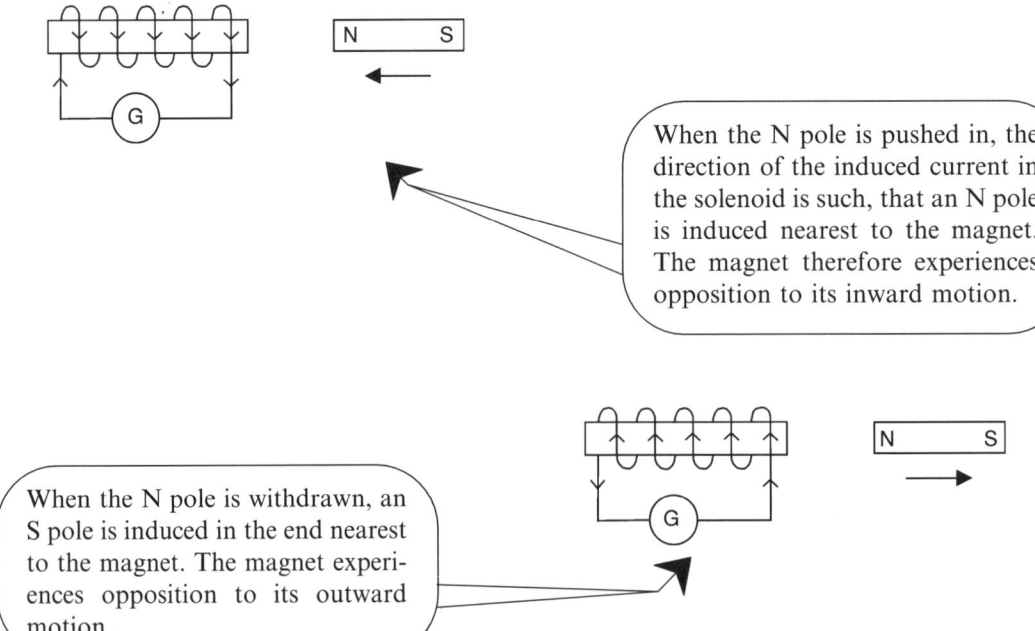

When the N pole is pushed in, the direction of the induced current in the solenoid is such, that an N pole is induced nearest to the magnet. The magnet therefore experiences opposition to its inward motion.

When the N pole is withdrawn, an S pole is induced in the end nearest to the magnet. The magnet experiences opposition to its outward motion.

3. Fleming's right hand dynamo rule

This rule is used to predict the direction of the induced current when a straight conductor moves in a magnetic field.

> • Put the forefinger, the middle finger and the thumb of the right hand at right angles to one another.
> • When the forefinger points in the direction of the magnetic field, the thumb in the direction of the thrust (movement), then the middle finger points in the direction of the induced conventional current.

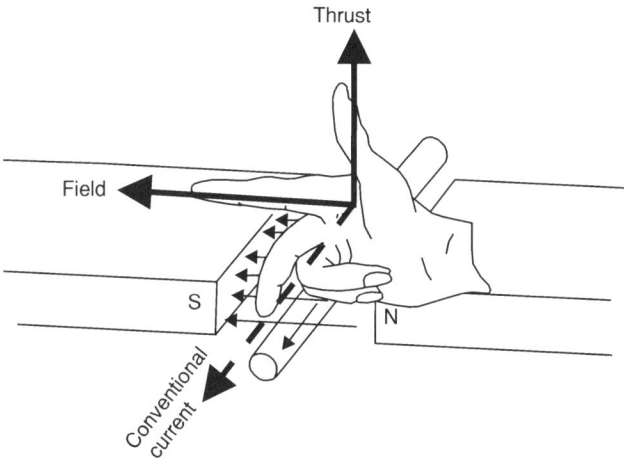

4. The simple alternating current dynamo

• The basic principle of the a.c. dynamo is **electromagnetic induction**.
• The dynamo relies on the principle that an **induced potential difference (emf) can be generated by a coil that turns in a magnetic field**.
• The **kinetic energy** of the conductor that turns in the magnetic field, changes into **electrical potential energy**.

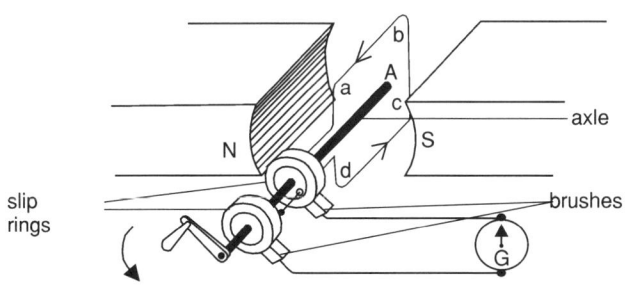

The simple a.c. dynamo

• The dynamo consists of a **rectangular coil** of wire rotating in a uniform magnetic field.
• The ends of the coil are connected to two **slip rings** mounted on the **axle**.
• Each slip ring makes contact with a **carbon or copper brush**.

- As the coil rotates, it moves across the **magnetic field**.
- There is a **change** in the **magnetic flux** linkage with the conductor and an **emf is induced** in the coil.
- The induced emf causes a current to flow if the circuit is closed.
- Fleming's right-hand rule can be used to determine the direction of the induced current.

Consider ab
- ab is moving downward; according to Fleming's dynamo rule the induced current moves from b to a.
- the current will be at its maximum when the coil is in the horizontal position and there is maximum rate of change of flux linkage.
- the current drops to zero when it turns into the vertical position.
- ab begins to move upwards. According to Fleming's dynamo rule the current now flows from a to b; increasing to maximum and then decreases to zero.

The induced current changes direction every 180° (half revolution) and is constantly changing in strength.

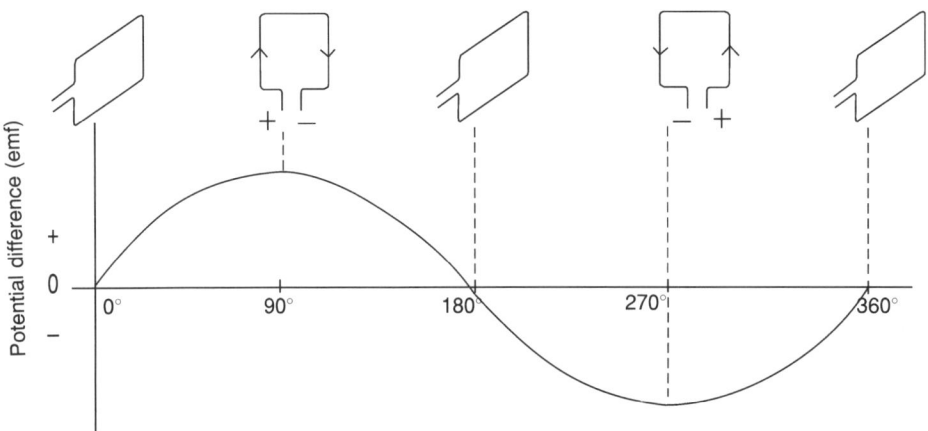

The graph shows the emf (voltage) induced in the coil of an alternating current dynamo during one complete revolution.

Ways of increasing the voltage of a simple dynamo

- Increase the number of turns on the coil
- Increase the speed of rotation
- Increase the strength of the magnetic field
- Winding the coil onto a soft iron core

5. Mutual magnetic induction

Thus far we have used the motion of a conductor in a magnetic field to produce the change in magnetic field linkage necessary for the induction of a current, according to Faraday's Law. We can also change field linkage by changing the current in one coil, the magnetic field of which is linked with another coil.

The change in current produces a change in the field linked with the second coil and a current is induced in the second coil. This is called Mutual Induction.

Experiment: To investigate the use of an electric current to induce current in another circuit.

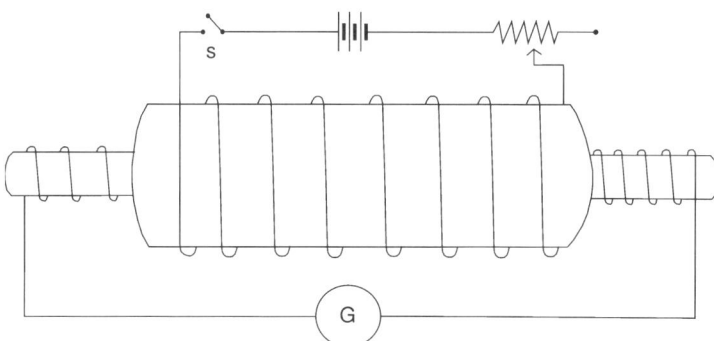

- Set up the circuit as shown in the sketch.
- The small solenoid (200 turns) is placed inside the bigger solenoid (100 turns).
- Close the switch and observe the galvanometer. Open the switch and watch the needle of the galvanometer.
- Place a soft iron core inside the smaller solenoid. Repeat the experiment.
- Use the rheostat to increase and decrease the magnitude of the current in the circuit of the big solenoid. Observe how the galvanometer reacts.

Results and deductions

- The circuit containing the source of electricity (cell) is called the **primary circuit**. The circuit in which a current (emf) is induced is called the **secondary circuit**.
- When the switch is closed the **current** in the primary circuit **increases from zero to maximum**. This **increasing current in the primary coil** produces a **changing magnetic field** around the secondary coil.
- This **change in the magnetic flux linkage** will cause a **current to be induced** in the secondary coil
- When a **constant current** flows in the **primary circuit**, the **flux linkage** with the secondary coil **stays the same** and **no current is induced**.
- **Current is induced in the secondary coil** only when there is an **increasing or decreasing current** in the **primary coil**.
- When an **iron core** is present, the induced current in the secondary coil is stronger. The iron core causes the flux density to increase (a stronger magnetic field). Thus, when the current is switched on or off, there is a **greater change in magnetic field linkage** and, therefore, a **stronger induced current**.

6. Alternating current transformer

The principle of **mutual induction** is applied in the design of the ac-**transformer**. When an alternating current (which is continually changing its direction) is passed through the primary coil, a **continually changing magnetic field induces** a **continually changing current in the secondary coil**.

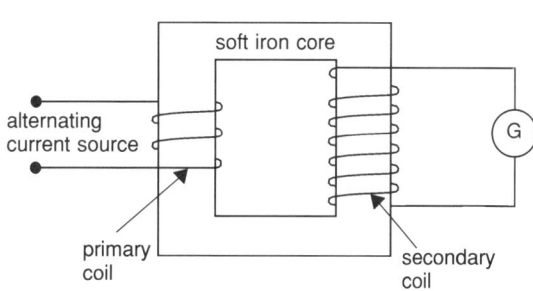

The transformer shown in the sketch at the bottom of the previous page consists of two coils of wire wound on the same soft iron core. The purpose of the soft iron core is to concentrate the magnetic fields produced.

- An **alternating current** is passed through the **primary coil**.
- This generates a **magnetic field** which is **continually changing**.
- The changing magnetic field passes round the soft iron core and through the **secondary coil**.
- The **changing magnetic field linkage with the secondary coil** causes **a current to be induced in the secondary coil**.

Transformers are divided into two types:

- **A step up transformer** has more turns of wire on the secondary coil than on the primary coil and **increases the potential difference** (emf).

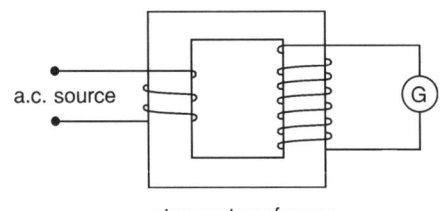

step up transformer

- **A step down transformer** has less turns of wire on the secondary coil than on the primary coil and **decreases the potential difference** (emf).

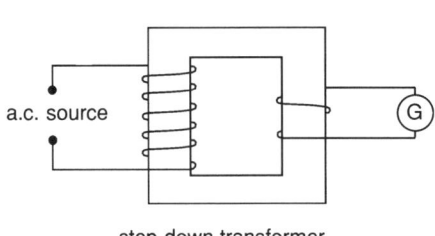

step down transformer

The following formula which combines the potential ratio and the turns ratio can be used to do calculations:

$$\frac{\text{Potential difference across the secondary coil}}{\text{Potential difference across the primary coil}} = \frac{\text{Number of turns on the secondary coil}}{\text{Number of turns on the primary coil}}$$

$\Rightarrow$ $$\frac{V_s}{V_p} = \frac{N_s}{N_p}$$

See part (a) of the worked out example on the next page as an application of this formula.

Higher Grade Only

Energy is lost in the transformer mostly because of the resistance of the wire of the coils. If the transformer is well designed, the energy losses are small, so:

$$\frac{\text{Power put into}}{\text{primary coil}} = \frac{\text{Power got out of}}{\text{secondary coil}}$$

$\Rightarrow$ $$V_p I_p = V_s I_s$$

114

where V_p = voltage in primary coil in volt

I_p = current in primary coil in ampere

V_s = voltage in secondary coil in volt

I_s = current in secondary coil in ampere

Example:

A transformer is required to step down a voltage of 220 V alternating current to 5 volt for an electric bell.

(a) If there are 1 100 turns on the primary coil, how many turns must be wound on the secondary coil?

(b) Assume no energy loss. Calculate the current in the primary coil of the transformer if the bell draws a current of 0,5 A.

Solution:

(a) $\dfrac{V_p}{V_s} = \dfrac{N_p}{N_s}$

$\dfrac{220}{5} = \dfrac{1100}{N_s}$

$N_s = \dfrac{1100 \times 5}{220}$

$= 25 \text{ turns}$

(b) $V_p I_p = V_s I_s$

$220 \times I_p = 5 \times 0,5$

$\therefore I_p = \dfrac{5 \times 0,5}{220}$

$= 0,011 \text{ A}$

QUESTIONS

Section A

I. *Various possibilities are suggested as answers to the following questions. Indicate the correct answers.*

1. A solenoid is connected to a galvanometer. A magnet is inserted into the solenoid and then kept stationary. The needle of the galvanometer will . . .
   A   gradually move to one side.
   B   move to one side and then fall back to the zero position.
   C   show no deviation.
   D   first move to the one side and then to the other side.

2. The conditions for a potential difference (emf) to be induced across the ends of a solenoid is that . . .
   A   there must be a change in magnetic flux linked with the solenoid
   B   both solenoid and magnet should be moved.
   C   the magnet must be moved while the solenoid remains stationary.
   D   there must be a second solenoid near the first one, in which a constant current flows.

3. A transformer can change . . .
   A   the voltage of both direct and alternating currents.
   B   direct current to alternating current.
   C   the voltage of an alternating current.
   D   alternating current to direct current.

4. In order to induce an emf (potential difference) in a coil by means of an electric current, one needs . . .
   A   a second coil, a battery and a magnet.
   B   a magnet and a switch.
   C   a second coil, a switch and a galvanometer.
   D   a second coil, a battery and a switch.

5. You are looking into a solenoid in which an electric current flows clockwise. The direction of the magnetic field inside the solenoid is . . .
   A   away from you.   C   to the right.
   B   towards you.   D   to the left.

6. Two appliances that use electromagnetic induction to function, are a . . .
   A   dynamo and electromagnet.   C   galvanometer and electromagnet.
   B   transformer and dynamo.   D   transformer and electric motor.

7. In a transformer the energy is transferred from the primary to the secondary coils by means of . . .
   A   electrical shockwaves.
   B   collisions between the molecules.
   C   the current in the secondary coil.
   D   changes in magnetic flux linkage.

8. Which one of the following components does not form part of both the alternating current dynamo and the direct current motor?
   A   Carbon brushes   C   Field magnets
   B   Splitring commutator   D   Insulated shaft

116

9. The primary coil of a transformer has 1 600 turns and is connected to a 220 V alternating current source. The secondary emf is 11 V. The number of turns on the secondary coil is . . .
   A   160.   B   32 000.   C   80.   D   800.

10. A transformer has 10 primary and 440 secondary windings. The ratio of the primary to secondary voltages will be . . .
    A   44:1.   B   1 100:1.   C   1:44.   D   1:1 100.

11. For a certain transformer $V_s < V_p$. Then . . .
    A $N_p > N_s$.   C $N_pV_p = N_sV_s$.
    B $N_p < N_s$.   D $N_pV_s < N_sV_p$.

12. The conventional current in the rectangular coil flows as indicated in the sketch. The coil will . . .
    A   rotate anticlockwise.
    B   rotate clockwise.
    C   as a whole move upwards.
    D   as a whole move downwards.

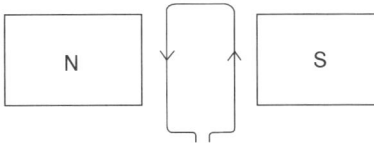

13. Which one of the following is an example of a step-up transformer?
    A   A ratio of $N_s : N_p = 20:200$.
    B   An emf of 220 V in the primary coil and 55 V in the secondary coil.
    C   200 Windings on the primary coil and 100 windings on the secondary coil.
    D   20 Windings on the primary coil and 100 windings on the secondary coil.

II. *Complete each of the following. Write down the missing word(s) next to the number of the question.*

1. A Galvanometer is a very sensitive ___1.1___ and it can indicate the presence of a ___1.2___ .

2. The dynamo always produces a ___2___ current.

3. The ___3.1___ energy of a coil moving in a magnetic field is transferred to ___3.2___ energy in the conductor.

4. The direction of the current produced by ___4.1___ induction may be determined by applying ___4.2___ Law. The direction of the induced current is always such that it ___4.3___ the inducing action.

5. When a ___5.1___ current flows in a circuit, no change in flux density occurs and ___5.2___ is induced in the secondary coil.

Section B

1. A galvanometer is connected to a solenoid. The N-pole of a magnet is inserted into the solenoid.

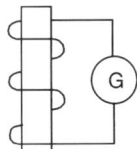

1.1 What will you observe on the galvanometer if the magnet's N-pole is

1.1.1 slowly pulled out of the solenoid.

1.1.2 quickly inserted into the solenoid.

1.1.3 slowly inserted into the solenoid.

1.1.4 held stationary inside the solenoid.

1.2 What is this phenomenon called?

1.3 Name two applications of this phenomenon.

2.1 What type of energy conversion occurs in the dynamo?

2.2 In which three ways can the induced emf be raised?

3. Two solenoids are placed next to each other with a soft iron core through both. One solenoid is connected to a battery and a switch and the other solenoid to a galvanometer.

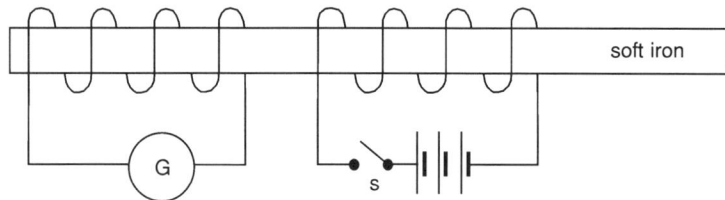

3.1 What will be observed when switch S is closed?

3.2 What will be observed when switch S is then opened?

3.3 How will the reading on the galvanometer be affected if the soft iron core is removed from both solenoids and steps 3.1 and 3.2 repeated.

3.4 What is this phenomenon in 3.1 called?

3.5 Name one apparatus in which this phenomenon is applied.

3.6 Explain how the current originates in the solenoid, which is connected to the galvanometer.

4. Use the accompanying sketch and explain how you would use Lenz's Law to determine the direction of the current induced in the coil when the S pole of a magnet is removed from the coil.

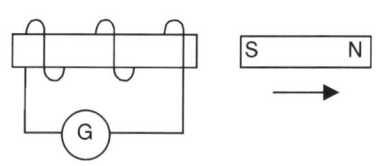

5. An electric bell operates on a 6 V alternating source. Only a 240 V alternating source is available. A transformer has 3 600 windings on the primary coil. Calculate the number of windings necessary on the secondary coil for the bell to function properly.

6.1 State Lenz's Law.

6.2 Use Lenz's Law to determine the direction of the current if the magnet is moved into the coil. State whether the current flows as shown at x or at y.

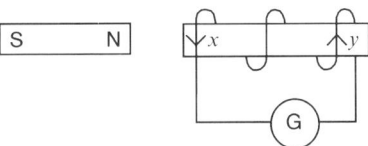

Additional for Higher Grade

7. The primary coil of the transformer in the sketch is connected to a 200 V alternating current source. The secondary coil is connected to an electric motor M. The maximum power output of the motor is 7,2 kW at a voltage of 10 V.

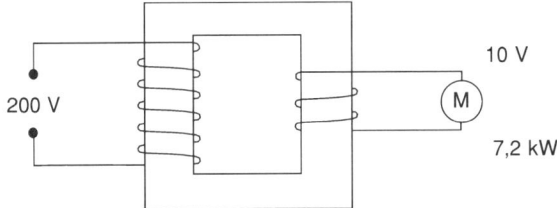

7.1 On which principle does the transformer function?

7.2 Why does a transformer need alternating current to function?

7.3 Can a transformer raise both potential difference and current? Motivate your answer.

7.4 Calculate the current in the primary coil.

7.5 Calculate the turns ratio.

7.6 What type of transformer is this?

8. The primary coil of the transformer has 6 turns while the secondary coil has 24 turns.

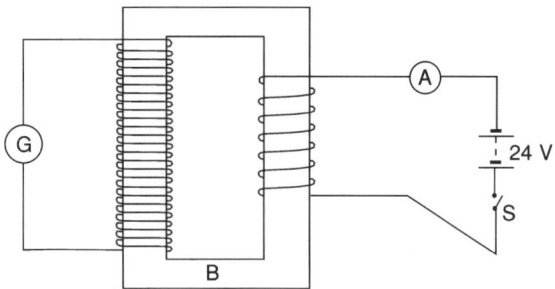

8.1 Switch S is closed for 2 minutes and then re-opened. State and explain what you will observe on the galvanometer.

The battery is now replaced by a 24 V alternating current source.

8.2 What will the ratio between the primary voltage and the secondary voltage be?

8.3 What will the current ratio be?

8.4 What is B called and what is its function?

8.5 If the ammeter registers a reading of 0,2 A when S is closed, what value will the galvanometer reading reach?

9. The number of turns on the secondary coil of a transformer is 40 times more than that on the primary coil. A potential difference of 200 V is applied across the primary coil and an alternating current of 2 A flows through the coil. Calculate

9.1 the secondary emf.

9.2 the effectiveness of the transformer if the current in the secondary coil is 0,02 A.

10. A current of 4,5 A flows through the primary coil of a transformer when connected to a 200 V source. The emf in the secondary coil is 15 V. Calculate

10.1 the current in the secondary coil.

10.2 the turns ratio of the transformer.

ANSWERS

Section A

I. 1. B   2. A 3. C   4. D   5. A   6. B   7. D
    8. B   9. C   10. C   11. A   12. B   13. D

II.   1.1 current detector   4.1 electromagnetic

      1.2 potential difference   4.2 Lenz

      2.   alternating   4.3 opposes

      3.1 kinetic   5.1 constant

      3.2 electrical potential   5.2 no current

Section B

1.1.1 The needle of the galvanometer moves to one side and then falls back to zero.

1.1.2 The needle moves far out to the opposite side and then falls back to zero.

1.1.3 The needle moves slightly to the same side as 1.1.2 and then falls back to zero.

1.1.4 No movement of the needle of the galvanometer.

1.2   Electromagnetic induction.

1.3   Dynamo; transformer.

2.1   Kinetic energy to electric potential energy.

2.2   • use stronger magnets
           • use more windings on the coil
           • rotate the coil faster.

3.1   The needle on the galvanometer moves to one side and then falls back to zero.

3.2   The needle moves to the other side and then falls back to zero.

3.3   There will be no movement of the needle. (There will be no magnetic flux linkage or at best very little.)

3.4   Mutual induction

3.5   Transformer.

3.6   • When S is closed the current increases from zero to a maximum.
           • This causes a changing magnetic field.
           • This changing magnetic field passes through the iron core to the secondary coil.
           • The changing magnetic flux linkage induces a current in the secondary coil.

4.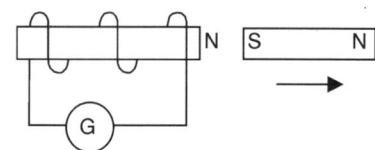

- When the S pole is withdrawn, an N pole will be induced at that end of the solenoid.
- According to Lenz's Law the induced current flows in such a direction that the magnetic field that it creates opposes the inducing action.

5.

$$\frac{N_s}{N_p} = \frac{V_s}{V_p}$$

$$\therefore \frac{N_s}{3600} = \frac{6}{240}$$

$$N_s = \frac{6 \times 3\,600}{240} = 90 \text{ windings.}$$

6.1 The induced current flows in such a direction that the magnetic effect it produces will oppose the inducing current.

6.2 Current as shown in y.

Additional for HG

7.1 Mutual induction.

7.2 For an emf to be induced in the secondary coil there must be a change in the magnetic flux linkage. This can be obtained by sending an alternating current through the primary coil. (Direct current will not cause a changing magnetic flux.)

7.3 The energy input per second in the primary coil can at its most be equal to the energy output per second in the secondary coil. (Law of the Preservation of Energy.)

$$\therefore \text{ power in secondary coil} = \text{power in primary coil}$$

$$\text{or } V_s I_s = V_p I_p$$

When V increases, I decreases in the same ratio.

7.4 Power in secondary coil = power in primary coil = 7 200 W

$$\text{or } V_p I_p = 7\,200 \text{ W}$$

$$\therefore I_p = \frac{7\,200}{V_p} = \frac{7\,200}{200} = 36 \text{ A.}$$

7.5 $\dfrac{N_p}{N_s} = \dfrac{V_p}{V_s} = \dfrac{200}{10} = 20{:}1$

7.6 Step down transformer.

8.1 • There will be a momentary deflection of the needle of the galvanometer.
• Then it returns to zero.
• When S is opened after 2 minutes there will again be a deflection of the needle in the opposite direction and it will return to zero again.
• Only when the switch is closed and opened there is a change in current which causes a change in the magnetic flux linked with the secondary coil.
• During the 2 minutes a direct current flows through the primary coil.
• There will be no emf induced across the secondary coil because there is no change in the magnetic flux linkage with the secondary coil.

8.2 $\dfrac{V_p}{V_s} = \dfrac{N_p}{N_s} = \dfrac{6}{24} = \dfrac{1}{4} = 1:4$

8.3 $I_p:I_s = 4:1$

8.4 Iron core
• This concentrates the magnetic field produced by the current in the primary coil.
• It also ensures a higher induced emf in the secondary coil.

8.5 $V_s\,I_s = V_p\,I_p$

$$I_s = \frac{V_p\,I_p}{V_s}$$

$$= \frac{24\,V \times 0,2\,A}{96\,V}$$

$$= 0,05\,A$$

or $\dfrac{I_p}{I_s} = \dfrac{4}{1} = \dfrac{0,2}{0,05}$

$\therefore I_s = 0,05\,A$

$\dfrac{V_p}{V_s} = \dfrac{1}{4} = \dfrac{24}{96}$

$\therefore V_s = 96\,V$

9.1 $N_s = 40\,N_p$
$V_p = 200\,V$
$I_p = 2\,A$

$\dfrac{V_p}{V_s} = \dfrac{N_p}{N_s}$

$\therefore \dfrac{200}{V_s} = \dfrac{N_p}{40\,N_p}$

$\therefore V_s = 200 \times 40 = 8\,000\,V$

9.2 $\dfrac{I_p}{I_s} = \dfrac{N_s}{N_p}$

$\dfrac{2}{I_s} = \dfrac{40\,N_p}{N_p}$

$\therefore I_s = \dfrac{2}{40} = 0,05\,A$

% effectiveness $= \dfrac{0,02}{0,05} \times 100 = 40\%$

10.1 $V_s I_s = V_p I_p$

$$I_s = \frac{V_p I_p}{V_s} = \frac{200 \times 4,5}{15} = 60 \text{ A}$$

10.2 $N_p : N_s = V_p : V_s$

$$= 200 : 15$$
$$= 40 : 3$$

8 Atomic Structure, the Periodic Table and Chemical Bonding

1. The history of the atom

Democritus

The Greek philosopher, **Democritus**, is the first person on record to suggest the idea that all matter is made up of tiny particles. These **particles** were called **atoms** (meaning indivisible)

Dalton

More than 2000 years passed before **John Dalton** investigated the idea of the atom again. Dalton had practical proof for his ideas and proposed the **billiard ball model** of the atom. He made the assumptions that:

- all matter consists of solid indivisible particles called atoms
- all atoms of the same element are identical
- atoms of different elements differ from one another
- atoms combine in whole number ratios with one another
- a combination of two or more atoms are called a compound.

Thomson

The **electron** was discovered by using cathode ray tubes to investigate the nature of electrical charges. To try and explain the electrical nature of matter **Thomson** established his **currant bun model** of the atom. The atom was described as a sphere, consisting of a solid, positively charged mass in which the negatively charged electrons were distributed regularly. The atom as a whole was neutral.

Rutherford

It soon became apparent that Thomson's model was not correct. **Ernest Rutherford** carried out his α-**particle-scattering experiment** to investigate the nature of the atom. He bombarded a thin sheet of gold foil with helium nuclei (α-particles) and found that most particles went straight through the gold foil, some were deflected through large angles and some of the particles almost reversed direction completely. From these observations Rutherford concluded that:

- most of the atom is **empty space**
- a **small nucleus** exists inside the empty space of the atom
- the nucleus is **positively charged**
- all the **mass** of the atom is practically concentrated in the **nucleus**
- the nucleus is surrounded by the **negatively charged electrons**.

Bohr

Niels Bohr located the **orbits** in which the electrons move around the nucleus. Each orbit can accommodate only a certain number of electrons and the energy of electrons differs from one orbit to another. Electrons radiate energy when they move from one stationary state to another of lower energy. This set the stage for the modern model of the atom called the **wave-mechanical** (or orbital) **model.**

2. Basic concepts of Chemistry

Matter

Matter can be regarded as anything that occupies space and possesses mass. Everything around us consists of matter which can be made up of three different types of substances:

- **Elements** – these are pure substances that cannot be broken down into simpler substances by chemical methods. Elements can be divided into two subgroups, metals and non-metals. Elements are in turn all made up of atoms.

The following table gives the differences between metals and non-metals:

Metals	Non-metals
• good conductors of electricity • good conductors of heat • malleable and ductile and can be worked into different shapes • shiny	• poor conductors of electricity • poor conductors of heat • brittle and cannot be beaten into different shapes like metals • not shiny

- **Compounds** – these are pure substances made up of elements that are chemically combined in fixed ratios. They can be broken down into simpler substances by chemical methods.

- **Mixtures** – these are impure combinations of two or more elements or compounds.

The following table gives the most important distinguishing properties of mixtures, compounds and elements.

Element	Compound	Mixture
• consist of one kind of atom	• composition is constant	• composition can vary and consists of two or more elements or compounds
• properties are unique to the element	• properties differ from those of the constituent substances	• properties are the same as those of the constituent substances
• cannot be separated by chemical or physical methods	• can be separated by chemical methods	• can be separated by physical methods
• Example: Cu, Na, F_2, etc.	• Example: H_2O, NH_3, etc.	• Example: $H_2 + O_2$

3. Atomic Structure

3.1 Atom

An atom consists of two main parts:
- a **nucleus** with **protons** and **neutrons** in it.
- a cloud of **electrons** surrounding the nucleus.

The **mass** of the atom is concentrated in the nucleus. The **protons** are positively charged and the **neutrons** have no charge at all. The electrons are negatively charged and have negligible mass.

3.2 Atomic number

An element is identified by its atomic number. **This is the number of protons in the nucleus of the atom**. All the atoms of the same element have the same number of protons in the nuclei of their atoms. The symbol for atomic number is the letter Z. The atomic number also gives the number of electrons in a neutral atom of the element.

3.3 Mass number

This is the total number of protons and neutrons in the nucleus of an atom. The symbol for mass number is the letter A.

A-Z = (number of protons + neutrons) − (number of protons)
    = number of neutrons in the nucleus

A very comprehensive way of **indicating an element** is:

Notation: $^A_Z X$
A: mass number
Z: atomic number
X: symbol of the element

> ## Example:
> Give the element, the number of protons, the number of neutrons and the number of electrons in $^{33}_{16}S$.
>
> ## Solution:
> The element is sulphur.
> There are 16 protons.
> There are 16 electrons.
> There are (33 − 16 = 17) neutrons.

3.4 Relative atomic mass

This is the **mass of an atom of an element relative to the mass of a carbon-12 atom**. The mass of a carbon-12 atom is exactly 12 units. The relative atomic mass of a sodium (Na) atom is 23, which simply means that it is 23 times heavier than the mass of one twelfth of a carbon-12 atom. The atomic masses are given on the Periodic Table of elements.

3.5 Relative formula mass

The relative formula mass of a compound uses the symbol M_r. It is calculated by adding the relative atomic masses of all of the elements that the compound consists of.

> **Example:**
>
> Calculate the relative formula mass of calcium carbonate ($CaCO_3$).
>
> **Solution:**
>
> $M_r(CaCO_3) = 40 + 12 + 3(16) = 100$.

3.6 Isotopes

Isotopes are atoms of the same element which have the same atomic number but different mass numbers.

> **Example:**
>
> $^{12}_{6}C$ and $^{14}_{6}C$ are isotopes of carbon because they each have 6 protons but one has 6 neutrons and the other one has 8 neutrons.

3.7 Ions

When an atom loses an electron it forms a **positive ion** called a **cation**. When an atom gains an electron it forms a **negative ion** called an **anion**.

4. The Periodic Table of the Elements

The Periodic Table lists all the elements in order of their **atomic number** i.e. the **number of protons in the nucleus**. In doing this we find that at certain intervals or **periods**, the properties of the elements are the same.

The **vertical columns** on the Periodic Table are known as **groups** and the **horizontal rows** are called **periods**.

- **Group I**-elements are known as the **alkali metals**.
- **Group II**-elements are known as the **alkali earth metals**.
- **Group VII**-elements are known as the **halogens**.
- **Group 0 or group VIII**-elements are known as the **noble gases**.

The middel block of elements without group numbers are called the **transition elements**.

The **properties of elements change gradually from left to right**, while the elements in a group are remarkably similar. In each group the properties of the elements differ in degree from top to bottom in each group.

Metals are located to the **left of the Table**, while the **non-metals** are situated to the **right**. The reactivity of metals decreases rapidly from left to right on the Periodic Table.

5. Electronic Structure

The electrons which surround the nucleus are not just randomly arranged, they can be found in energy levels which radiate from the nucleus. They are numbered 1, 2, 3, etc. with the one closest to the nucleus being number 1.

5.1 Orbitals

Within the energy levels there are sub-levels called **orbitals**. **An orbital is a space in which the probability (chance) of finding an electron is the greatest**.

Table 8.1 The Periodic Table of Elements

KEY

29	Atomic number
Cu	Symbol
63,5	Relative atomic mass (Approximately)

I	II											III	IV	V	VI	VII	0
1 **H** 1																	2 **He** 4
3 **Li** 7	4 **Be** 9											5 **B** 11	6 **C** 12	7 **N** 14	8 **O** 16	9 **F** 19	10 **Ne** 20
11 **Na** 23	12 **Mg** 24											13 **Aℓ** 27	14 **Si** 28	15 **P** 31	16 **S** 32	17 **Cℓ** 35,5	18 **Ar** 40
19 **K** 39	20 **Ca** 40	21 **Sc** 45	22 **Ti** 48	23 **V** 51	24 **Cr** 52	25 **Mn** 55	26 **Fe** 56	27 **Co** 59	28 **Ni** 59	29 **Cu** 63,5	30 **Zn** 65	31 **Ga** 70	32 **Ge** 73	33 **As** 75	34 **Se** 79	35 **Br** 80	36 **Kr** 84
37 **Rb** 86	38 **Sr** 88	39 **Y** 89	40 **Zr** 91	41 **Nb** 92	42 **Mo** 96	43 **Tc**	44 **Ru** 101	45 **Rh** 103	46 **Pd** 106	47 **Ag** 108	48 **Cd** 112	49 **In** 115	50 **Sn** 119	51 **Sb** 122	52 **Te** 128	53 **I** 127	54 **Xe** 131
55 **Cs** 133	56 **Ba** 137	57 **La** 139	72 **Hf** 179	73 **Ta** 181	74 **W** 184	75 **Re** 186	76 **Os** 190	77 **Ir** 192	78 **Pt** 195	79 **Au** 197	80 **Hg** 201	81 **Tl** 204	82 **Pb** 207	83 **Bi** 209	84 **Po**	85 **At**	86 **Rn**
87 **Fr**	88 **Ra** 226	89 **Ac**															

58 **Ce** 140	59 **Pr** 141	60 **Nd** 144	61 **Pm**	62 **Sm** 150	63 **Eu** 152	64 **Gd** 157	65 **Tb** 159	66 **Dy** 163	67 **Ho** 165	68 **Er** 167	69 **Tm** 169	70 **Yb** 173	71 **Lu** 175
90 **Th** 232	91 **Pa**	92 **U** 238	93 **Np**	94 **Pu**	95 **Am**	96 **Cm**	97 **Bk**	98 **Cf**	99 **Es**	100 **Fm**	101 **Md**	102 **No**	103 **Lr**

There are two different types of orbitals that we need to consider, **namely s and p orbitals**.

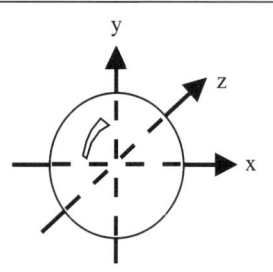

The shape of the s-orbitals are spherical and they occur singly. The x, y and z axes shown here are at right angles to each other and has no real significance here, but look on the right!!

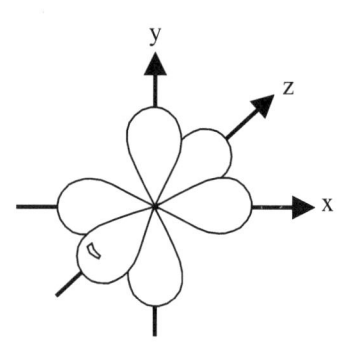

The p-orbitals consist of two lobes forming a figure 8 (eight). They appear in groups of three. The three orbitals are at right angles to each other.

There is an s-orbital in every energy level, but p-orbitals only occur from the second energy level onwards. The orbitals are indicated as follows: A numeral in front to indicate the energy level and then the letter to indicate the type of orbital, e.g. 3s refers to the s-orbital in the third energy level. Orbitals are represented by small circles.

Energy level	Orbitals	Representation
4	{ 4s	O 4s
3	(3p	O O O 3p
	(3s	O 3s
2	(2p	O O O 2p
	(2s	O 2s
1	{ 1s	O 1s

Energy level diagram

- The order in which the orbitals are filled, are 1s, 2s, 2p, 3s, 3p, 4s.
- The electrons are represented by arrows pointing in opposite directions ↑↓ to indicate that the electrons have opposite spin (i.e. clockwise or anti-clockwise).

When working out the electron structure the following rules apply:

- An **electron must** always **enter** the **lowest available energy level**.
- In an energy level **s-orbitals will be filled before the p-orbitals**.
- There may be a **maximum of two electrons in any one orbital**, and they must have opposite spin.
- Where there are **equivalent orbitals** available (e.g. p-orbitals) each orbital must have one electron before any one can have two.

Example:

Draw an energy level diagram to show the arrangement of electrons in $^{31}_{15}P$.

Solution: 4s   O

3p ⦶ ⦶ ⦶

3s   ⦿

2p ⦿ ⦿ ⦿

2s   ⦿

1s   ⦿

5.2 Electron configurations

The information of the example above can be given in summarised form as an **electron configuration**.

$^{31}_{15}P$ $1s^2\ 2s^22p^6\ 3s^23p^3$

5.3 Periodicity of electronic structure

Elements in the same group will have the same outer electron structure, despite having different numbers of electrons.

Look at the electron configurations of the following elements of Group I.

$_3Li$	$1s^2\ 2s^1$	These elements all have their outermost electron in an s-orbital.
$_{11}Na$	$1s^2\ 2s^22p^6\ 3s^1$	There is **one** outermost electron.
$_{19}K$	$1s^2\ 2s^22p^6\ 3s^23p^6\ 4s^1$	

Look at the elements of Group VII.

$_9F$	$1s^2\ 2s^22p^5$	Each element has seven electrons in the highest occupied energy level.
$_{17}C\ell$	$1s^2\ 2s^22p^6\ 3s^23p^5$	

This explains why elements in the same group tend to have similar properties and reactions.

5.4 Valence electrons

The electrons in the highest occupied energy level of an atom are known as the valence electrons.

Example:

$_{12}$Mg

3p O O O
3s ⓪

2p ⓪ ⓪ ⓪
2s ⓪

1S ⓪

$_7$N

3p O O O
3s O

2p ⓪ ⓪ ⓪
2s ⓪

1s ⓪

Mg has 2 valence electrons and is in Group II.
N has 5 valence electrons and is in Group V.

Elements are arranged on the Periodic Table in order of their atomic numbers and in groups with the same number of valence electrons. The group numbers correspond with the number of valence electrons.

5.5 Valency

This is the number of electrons which an atom must either gain or lose or share in order to achieve an outer electron configuration which is the same as that of the nearest noble gas.

Elements in group I, II, III and IV have a valency equal to the group number.
Elements in group V, VI and VII have a valency equal to (8 – group number).

Example:

$_{13}$Aℓ

⓪ O O 3p
⓪   3s

⓪ ⓪ ⓪ 2p
⓪   2s

⓪   1s

Aℓ is in group III
Aℓ has 3 valence electrons
Aℓ has a valency of 3

$_{16}$S

⓪ ⓪ ⓪ 3p
⓪   3s

⓪ ⓪ ⓪ 2p
⓪   2s

⓪   1s

S is in group VI
S has 6 valence electrons
S has a valency of 8 – 6 = 2.

6. Chemical Bonding

Chemical compounds differ with respect to their chemical and physical properties. Chemical compounds can be divided into two categories, namely **covalent** and **ionic compounds**.

A **chemical bond** is a force which holds the atoms together to form a single unit. When two atoms form a bond it is the **valence electrons** which interact.

The **noble gases** (Group 0) are very unreactive and form no compounds. The electronic configuration of these elements are characterised by **full s and p orbitals**, an electronic configuration that we call the noble gas or **octet structure**. This electronic configuration is very stable. All other elements try to achieve this same outer electron configuration when they share, lose or gain electrons.

In short: Only the **atoms of the noble gases are inactive** because they have the **octet structures in their outer energy levels.** All the other atoms on the Periodic Table are "unsatisfied" and they are willing to share, lose or gain electrons in order to acquire an octet electron structure in their outer energy levels.

6.1 Covalent bonding

This type of bond occurs when **electrons are shared**. It occurs between **elements which are non-**

132

metals. When two atoms each supply an electron for sharing, a single covalent bond is formed. **Covalent bonding forms compounds called molecules**.

- An **orbital** containing **one electron** is called a **half-filled orbital**.
- When **half-filled orbitals** of two atoms **overlap** a covalent bond is formed.
- A **shared electron pair** is formed.
- The shared electron pair must have **opposite spin** and will be accommodated in a common orbital.

Example: Hydrogen (H_2)

When two hydrogen atoms approach each other, their two half-filled 1s orbitals overlap. The two shared electrons must have opposite spin and now form an electron pair. This electron pair revolves around and between the two nuclei. Thus, they can be said to belong to both atoms.

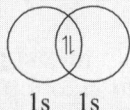

1s   1s

Overlapping of the orbitals in a molecule of hydrogen

This formation of the H_2-molecule can also be illustrated by means of Lewis structures.

$H° + {}_xH → H{}_x^°H$

Example: Water (H_2O)
Consider the electronic configurations of the oxygen and the hydrogen atoms.

⥮ ⥮ ⥮ 2p		
⥮ 2s		
⥮ 1s	⥮ 1s	
Hydrogen	Oxygen	

Lewis diagrams:

- Write the symbol of the element and indicate the valence electrons only by means of dots or crosses around the symbol.
- Indicate whether the electrons are paired or unpaired.
- In molecules the shared electrons are indicated by placing them between the elements.
- Once drawn, each atom in the molecule must have a completed outer energy level.
- Count the shared elections in each atom in each bond.

The oxygen has two unpaired electrons while the hydrogen atom has only one. Covalent bonding occurs when the 1s orbital of two hydrogen atoms overlap with each of the two half-filled 2p orbitals of the oxygen.

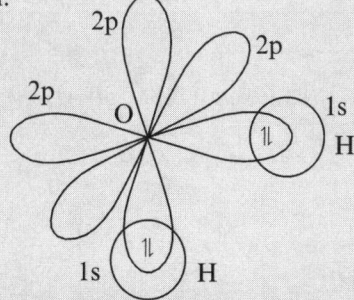

Lewis structures can also be used to illustrate the formation of the H_2O-molecule:

$$H^x + H^x + {}^{oo}_{o}\!O{}_{o}^{} → {}^{oo}_{o}\!O{}^x_{ox} H$$
$$H$$

Examples where only Lewis structures are used:

(a) HF: $H_x^x F_x^{x}$

(c) F_2: $_x^x F_x^x F_o^o$

(b) NH_3: $H_o^o N_x^o H$
$\qquad\quad H$

(d) O_2: $_x O_x^{xx} O_o^{oo}$

Thus, when chemical formulae are written, the correct number of atoms of each element participating in the molecule is obtained by determining how many electrons each atom has to share.

6.2 Ionic bonding

This type of bond generally occurs between a **metal** and **non-metal** and is characterized by the **transfer of electrons**.

- **Atoms of metals** easily **donate** an **electron** to form a **positive ion**
- Atoms of **non-metals** easily **accept** an **electron** to form a **negative ion**
- The metal atom has **transferred an electron** to the atom of the non-metal
- The **positive** and **negative ions attract each other** by electrostatic forces because they have opposite charges.
- This attracting force forms the **ionic bond**.

The formation of the ionic bonding between sodium and chlorine can be represented by three steps:

- The formation of the positive ion:

$$Na^\circ \rightarrow [Na]^+ + e^-$$

- The formation of the negative ion:

$$_x^x Cl_x^x + e^- \rightarrow [_x^x Cl_o^x]^-$$

- The positive and the negative ions pack together in an orderly way to form a crystal lattice:

$$[Na]^+ + [_x^x Cl_o^x]^- \rightarrow [Na]^+[_x^x Cl_o^x]^-$$
$$\rightarrow NaCl(s)$$

The charges on the ions are very easy to determine:

- Metals of Group I, II and III – the charge on the ion is indicated by the group number.
- Non-metals of Group V, VI and VII – the charge on the ion is the difference between eight and the number of the group.

Example:

Use Lewis structures to represent the formation of bonds between calcium and fluorine.

Solution:

Formation of the positive ion:

$$Ca_o^o \rightarrow [Ca]^{2+} + 2e^-$$

Formation of the negative ion:

$$2_x^x F_x^x + 2e^- \rightarrow 2[_x^x F_o^x]^-$$

Packing of the ions in a crystal lattice:

$$[Ca]^{2+} + 2[_x^x F_o^x]^- \rightarrow Ca^{2+}(F^-)_2 \rightarrow CaF_2(s)$$

134

Where elements combine in this way, the result is not a molecule because single units of ionic compounds do not exist. **Ionic compounds form crystal lattices made up of alternating positive and negative ions**.

Polyatomic (compound) ions

These consist of two or more atoms. They behave as single ions and they can be combined with other ions in the same way.

Below is a list showing the valencies and ionic charges of elements and some of these polyatomic ions:

One bond	Two bonds	Three bonds	Four bonds
H^+ hydrogen Na^+ sodium K^+ potassium Ag^+ silver Cu^+ copper(I) NH_4^+ ammonium	Ba^{2+} barium Ca^{2+} calcium Cu^{2+} copper(II) Fe^{2+} iron(II) Pb^{2+} lead(II) Mg^{2+} magnesium Zn^{2+} zinc	Fe^{3+} iron(III) Al^{3+} aluminium	
			C carbon
Cl^- chlorine Br^- bromine I^- iodine NO_2^- nitrite NO_3^- nitrate ClO_3^- chlorate OH^- hydroxide HSO_4^- hydrogen- sulphate HCO_3^- hydrogen carbonate	O^{2-} oxygen S^{2-} sulphur SO_4^{2-} sulphate CO_3^{2-} carbonate	PO_4^{3-} phosphate	

Table 8.2 Valencies of elements and polyatomic ions

7. Writing chemical formulae

When the name of a compound is given, its formula can be written with the help of the following rules:

• Write down the symbols of the bonding elements and/or the compound ions. The symbol of the metal or positive ion is written on the left hand side of the formula.

• Determine the valencies of the bonding elements and/or compound ions.

• Write down the valencies of the different elements and ions above the symbols.

• Determine how many atoms of each element are required for the total valency of the atoms of the first element to be equal to the total valency of the atoms of the second element.

Examples:

(a) **lead(IV) oxide**
- Pb   O
- valency of lead = 4
- valency of oxygen = 8 – 6 = 2
- Pb$_4$ O$_2$
- 2 atoms of oxygen are required.
  The formula is PbO$_2$

Lead(IV) oxide indicates that the valency of the lead ion is +4. The valency table shows the normal valency of an element. Some metal ions, however, sometimes acquire different valencies and then the valency is indicated in the formulae by means of Roman figures.

(b) **aluminium sulphate**
- Aℓ   SO$_4$
- valency of aluminium = 3
- valency of SO$_4$ = 2
- Aℓ$_3$ SO$_4$$_2$
- 2 Aℓ atoms are required, because the total valency is 2 × 3 = 6
  3SO$_4$ ions are required, because the total ionic charge is 3 × 2 = 6
  The formula is Aℓ$_2$(SO$_4$)$_3$

Note: Brackets must be placed around the compound ion if more than one type of atom appear in the compound ion.

8. Chemical equations

The following rules will help you to balance equations:
- Write down the correct formulae of the reactants and the products.
- Determine how many atoms of each element occur on either side of the arrow.
- Add reactants or products to the left and right side of the arrow to balance the number of atoms of each element.
- Never change the formulae of reactants or products in order to balance the equation; just add balancing figures in front of a formula where necessary.

Example:

sodium + water → sodium hydroxide + hydrogen

$$Na + H_2O \rightarrow NaOH + H_2$$

Balance the hydrogen atoms by adding another H_2O on the left side and another NaOH on the right side.

$$Na + 2H_2O \rightarrow 2NaOH + H_2$$

To balance the Na atoms, add one more Na on the left side.

$$2Na + 2H_2O \rightarrow 2NaOH + H_2$$

9. Spontaneous, non-spontaneous, exothermic and endothermic reactions

- A reaction that does not start by itself is called a **non-spontaneous** reaction.
  Example: A candle never burst into flame by itself. It needs energy to start burning.
- A reaction that begins by itself is called a **spontaneous reaction**.
  Example: Phosphorous bursts into flame by itself when placed in oxygen and burns vigorously.
- An **endothermic reaction** absorbs energy continuously while it takes place.
  Example: NH_4NO_3 dissolves spontaneously in water, but the temperature of the water drops continuously.
- An **exothermic reaction** liberates energy continuously while it takes place.
  Example: Hydrochloric acid dissolves spontaneously in water and the temperature rises continuously while it dissolves.

136

QUESTIONS

Section A

*Various possibilities are suggested as answers to the following questions.
Indicate the correct answer.*

1. The elements of Group O are known as . . .
A	earth metals.	C	halides.
B	alkali-earth metals.	D	noble gases.

2. The fact that electrons of all substances have the same charge and mass was proved by . . .
A	Thomson.	C	Rutherford.
B	Dalton.	D	Democritus.

3. Alpha (α)-particles are . . .
A	neutrons.	C	nucleons.
B	protons.	D	protons and neutrons.

4. The atomic number of an element is determined by the number of . . .
A	protons + neutrons.	C	nucleons.
B	protons + neutrons + electrons.	D	protons.

5. The elements on the Periodic Table are arranged according to their . . .
A	physical properties.	C	number of protons in the nucleus.
B	chemical properties.	D	number of protons and neutrons.

6. Consider the isotope $^{25}_{12}X$. A nucleus of this atom contains . . .
A	12 neutrons and 13 protons.	C	12 protons and 13 neutrons.
B	12 protons and 25 neutrons.	D	12 neutrons and 25 protons.

7. When all the electrons of an atom appear in the first main energy level, the element is in . . .
A	group I.	C	group VII.
B	group II.	D	period 1.

8. When a chlorine atom forms an ion, the latter will have the same number of electrons as . . .
A	Ar.	C	Ca.
B	O^{2-}.	D	Na^{+}.

9. The relative formula mass of aluminium hydroxide is . . .
A	99.	C	27.
B	61.	D	78.

10. An electron will always attempt to enter into . . .
A   the energy level with the highest energy.
B   the most stable energy condition.
C   the orbital with the lowest energy.
D   the orbital with the highest energy.

11. Which of the following represents the correct electron configuration for $^{19}_{9}F$?
A	$1s^1\,2s^2 2p^6$	C	$1s^2\,2s^1 2p^6$
B	$1s^2\,2s^2 2p^5$	D	$1s^2\,2s^2 2p^6\,3s^2 3p^5 4s^1$

An atom of a certain element has the following electron structure:

Use this information to answer questions 12, 13, 14 and 15.

12. The atomic number of this element is . . .

    A   5.   C   12.

    B   15.   D   10.

13. The mass number of this atom is . . .

    A   6.   C   30.

    B   15.   D   indeterminable due to insufficient information.

14. The group in which this element occurs is . . .

    A   II.   C   IV.

    B   III.   D   V.

15. The element is a . . .

    A   metal.   C   transition element.

    B   non-metal.   D   noble gas.

16. Which of the following statements is correct when two or more atoms combine?

    I   The elements are kept together by electrostatic forces.

    II   The energy of the compound is lower than that of the seperate atoms.

    III   Atoms combine to obtain a more stable electron distribution.

    A   I and II   C   II and III

    B   I and III   D   I, II and III

17. In most cases an atom strives to have an outer energy level with . . .

    A   10 electrons   C   6 electrons

    B   8 electrons   D   4 electrons

18. A characteristic of covalent bonding is the formation of . . .

    A   polar molecules.   C   metals.

    B   crystal lattices.   D   molecules.

19. Between which pair of atoms will the formation of a covalent bond be the greatest possiblity?

    A   H and O   C   Ca and Ca

    B   Mg and $C\ell$ D   K and $C\ell$

20. In which of the following molecules does a s-orbital overlap with a p-orbital?

    A F_2 C CO_2

    B Br_2 D H_2O

21. When an atom gains an electron, it forms a . . .

    A   anion.   C   molecule.

    B   cation.   D   proton.

22. In which one of the following substances is there no covalent bonding?

    A   Magnesium   C   Hydrogen

    B   Chlorine   D   Bromine

23. In an ionic bond . . .

    A   the metal atoms attract the electrons.   C   the metal becomes the negative ion.

    B   the transfer of electrons occurs.   D   the non-metal becomes the cation.

24. The probability for an ionic bond to be formed is the greatest between . . .

    A   C and O.   C   F and F.

    B   Mg and Mg.   D   Ca and $C\ell$.

The electron distribution of an element X is as follows:

$$_8X: \ 1s^2 \ 2s^2 2p^4$$

Use this information to answer questions 25 to 28.

25. Which one of the following will represent an ion of X?

    A X^{2-} C X^+

    B X^- D X^{2+}

26. The charge that X obtains, can be represented by . . .

    A $X + 2e^- \rightarrow X^{2-}$.   C $X + e^- \rightarrow X^+$.

    B $X \rightarrow X^{2-} + 2e^-$.   D $X + 2e^- \rightarrow X^{2+}$.

27. The element is a . . .

    A   non-metal and belongs to Group VIII.   C   non-metal and belongs to Group VII.

    B   non-metal and belongs to Group VI.   D   metal and belongs to Group II.

28. If the element X reacts with Ca, it can be represented by . . .

    A $X^{2-} + Ca \rightarrow CaX_2$.   C $X^{2-} + Ca^{2+} \rightarrow CaX$.

    B $X^{2-} + Ca^{2+} \rightarrow Ca_2X_2$.   D $X^{2-} + Ca^{2-} \rightarrow CaX$.

29. The valency of an element is the number of electrons which an atom can . . .

    A   take up.   C   give off.

    B   share.   D   take up, give off or share.

30. The normal valency and ionic charge of oxygen are respectively . . .

    A   2 and $+2$.   C   6 and –6.

    B   2 and –2.   D   6 and $+6$.

31. Which of the following statements is **not true** for ionic crystals?

    A   Molecules consisting of atoms are formed.

    B   They conduct an electrical current in the liquid state.

    C   Electrons are transformed from a metal to a non-metal.

    D   The crystals are hard and brittle.

32. Which of the following equations does not apply to the formation of magnesium chloride?

    A $Mg \rightarrow Mg^{2+} + 2e^-$ C $Mg + C\ell_2 \rightarrow MgC\ell_2$

    B $\overset{x\,x}{\underset{x\,x}{_xC\ell^x}} + e^- \rightarrow \overset{x\,x}{\underset{x\,x}{_xC\ell^{x-}_{\circ}}}$ D $C\ell_2 - 2e^- \rightarrow 2 \ \overset{\circ\,\circ}{\underset{\circ\,\circ}{_{\circ}C\ell^x_x}}$

33. Covalent bonding . . .

    A   makes a compound a good conductor of electricity.

    B   makes a compound soluble.

    C   is the result of electron sharing.

    D   forms between metals and non-metals.

The following represent the valency electron distribution for A and B:

A ⭕⭕⭕ p   B ⓪⓪⓪ p
    ⓪ s   ⓪ s

Use the above information to answer questions 34, 35 and 36.

34. The number of valence electrons and the normal valency for B are respectively . . .
   A   1 and 1.   C   7 and 1.
   B   7 and 7.   D   5 and 1.

35. The bond between these two atoms is a . . .
   A   pure covalent bond.   C   double covalent bond.
   B   ionic bond.   D   single covalent bond.

36. The formula of the compound formed when the two atoms combine, is . . .
   A A_2B_7.   C AB_2.
   B A_2B.   D A_7B_2.

37. The formula $SO_3{}^{2-}$ represents . . .
   A   a sulphide ion.   C   a molecule.
   B   a sulphite ion.   D   a sulphate ion.

38. The forming of an aluminium ion is represented by . . .
   A $A\ell \rightarrow A\ell^{3+} + 3e^-$.   C $A\ell \rightarrow A\ell^{2+} + 2e^-$.
   B $A\ell \rightarrow A\ell^{+} + e^-$.   D $A\ell + 3e^- \rightarrow A\ell^{3-}$.

39. The bonding in a molecule of hydrogen bromide is . . .
   A   covalent as there is a transfer of electrons.
   B   covalent as there is a sharing of electrons.
   C   ionic as there is a transfer of electrons.
   D   ionic as there is a sharing of electrons.

40. The chemical behaviour of an element is determined by the number and arrangement of its . . .
   A   atoms.   C   protons.
   B   neutrons.   D   electrons.

Additional for the Higher Grade

41. A certain element has a mass number of $(2x + 18)$ where x is its atomic number. The number of neutrons in this atom is . . .
   A $3x + 18$.   C $x + 18$.
   B x.   D $2x + 18$.

42. A hydrogen ion with a positive charge is a . . .
   A   hydrogen atom with a negative charge.
   B   electrically charged hydrogen molecule.
   C   proton.
   D   hydrogen atom with an extra electron.

43. A symbol that can represent an isotope of oxygen, is . . .

   A $^{16}_{8}X$.   C $^{8}_{8}X$.

   B $^{8}_{5}X$.   D $^{17}_{10}X$.

44. In which of the following will all the particles have the same number of electrons?

A   Li, Na and K   C Cl^-, Ar and Na^+
B F^-, Na^+ and Ne   D   He, Ne and F^-

45. Which of the following formulae for compounds of metal M is not correct?

A MCl C $M(NO_3)_2$
B MO D MCO_3

46. Which of the following formulae is not correct for compounds of element E?

A Al_2E_3 C K_2E
B $(NH_4)_2E$ D H_2E_3

47. The normal valency of metal M in the formula M_2O_3 is . . .

A $+2$.   C   0.
B -2.   D   3.

48. The chromate ion, $CrO_4{}^{2-}$, combines with lead(IV) to form a compound with the formula . . .

A Pb_4CrO_4.   C $PbCrO_4$.
B $Pb(CrO_4)_2$ D Pb_2CrO_4.

49. Which of the following electron structures is incorrect?

A   H ⦂N⦂ H   C   ⦂O⦂H
         H   H

B   H⦂F⦂   D   ⦂N⦂ N⦂

Section B

1.1   Name two suppositions that Bohr had made in his model of the atom.

1.2   Name two shortcomings of this model.

2.   What was Rutherford able to conclude from his alpha (α) particle scattering experiment?

3.   Phosphorous has 3 electrons in the 3p sub **energy level** in its **ground state**.

3.1   What is meant by the bold typed terms?

3.2   Use the arrow-in-the-circle method to show the electron structure for phosphorous.

3.3   In which energy level of this element do we find the electrons with the most energy?

3.4   How many valence electrons does this atom have?

3.5   What is the maximum number of electrons that can go into a 3p orbital?

3.6   Are the three electrons in the 3p energy level of phosphorous paired or not? Give a reason for your answer.

3.7   Describe the form of a p orbital.

3.8   Name two differences between a 2p and a 3p orbital.

3.9   Name two differences between a 3s and a 3p orbital.

4. Complete the table below. All the atoms are neutral.

Atom (element)	Symbol	Atomic number	Mass number	Number of protons	Number of neutrons	Number of valence electrons	Normal valency	Ionic charge
	$^{19}_{9}\text{F}$							
		19	39					
Argon			40					
			24	12				
	$^{27}_{13}\text{A}\ell$							
				16	16			
Nitrogen					7			
	$^{14}_{6}\text{C}$							

5.1 Distinguish between the number of orbitals in the first and second energy levels. Name the orbitals.

5.2 What is the maximum number of electrons in the second energy level?

5.3 Draw energy level diagrams for $^{11}_{5}\text{B}$, $^{16}_{8}\text{O}$ and $^{40}_{20}\text{Ca}$.

6. A certain element X lies in group VI on the Periodic Table.

   6.1 What is the normal valency of this element?

   6.2 Give the ionic charge of this element.

   6.3 How many half-filled orbitals does the element have?

   6.4 Which orbitals are the half-filled ones?

   6.5 Write down a chemical equation to show the formation of an ion of this element.

   6.6 Write down the formulae should

   6.6.1 X and K (potassium) combine.

   6.6.2 X and Aℓ (aluminium) combine.

   6.7 Name the type of chemical bond formed in question 6.6.1 and 6.6.2.

7. An element has the electron configuration $1s^2\ 2s^2 2p^6\ 3s^2 3p^1$.

   7.1 In which group of the Periodic Table is it found?

   7.2 In which period of the Periodic Table is it found?

   8.1 What is characteristic of a Lewis diagram?

   8.2 Give the Lewis diagrams for Li, N and Cℓ.

   8.3 Use Lewis structures to show how the bonding between the following elements take place:

   8.3.1 Na and F

   8.3.2 H and S

   8.3.3 F and F

8.3.4 Mg and Cℓ

8.3.5 O and O

8.4 Indicate if the bond is ionic or covalent.

9. Consider the following and then answer the questions:

$^{19}_{9}X^-$; $^{20}_{10}Y$; $^{23}_{11}R^+$; $^{27}_{13}S^{3+}$

9.1 Name any similarities between X and Y.

9.2 Do R and X have identical chemical properties? Give a reason for your answer.

9.3 Name any differences between X and Y.

9.4 Which of these are anions?

9.5 Which of these have a noble gas structure?

9.6 Which elements are represented by Y and S?

10. Identify the type of bonding which occurs in each of the following substances:

10.1 KF

10.2 H_2O

10.3 chlorine gas

10.4 CaF_2

10.5 CO_2

10.6 CaO

10.7 magnesium hydroxide

10.8 lithium oxide

10.9 H_2S

10.10 calcium phosphate

11. Write chemical formulae for:

11.1 chlorine gas

11.2 magnesium hydroxide

11.3 lithium oxide

11.4 calcium phosphate

11.5 aluminium sulphate

11.6 sulphuric acid

11.7 sodium nitrate

11.8 potassium fluoride

11.9 sodium carbonate

11.10 calcium nitrate

12. Write down the names of the following compounds:

12.1 K_2SO_4

12.2 $CaCO_3$

12.3 $Aℓ(OH)_3$

12.4 $Zn(NO_3)_2$

12.5 FeO

12.6 KOH

12.7 $NaHCO_3$

12.8 H_2S

13. Indicate whether the following reactions are spontaneous or non-spontaneous and whether they are exothermic or endothermic:

13.1 The lighting of a match.

13.2 The melting of an ice block at room temperature.

13.3 An explosion of a bomb.

13.4 Sulphuric acid is added to water and the temperature increases.

14. Hydrogen burns in air and forms water.

    14.1 Write a balanced chemical equation for this reaction.

    14.2 What type of chemical bond would you find in the water molecule?

    14.3 Use Lewis structures to illustrate the bonding in this molecule.

    14.4 Draw a labelled sketch to illustrate the overlapping orbitals in the water molecule.

    15.1 Explain what is meant by

        15.1.1 atomic number

        15.1.2 isotope

        15.1.3 valency.

    15.2 Consider the following atom and answer the questions: $^{23}_{11}X$

        15.2.1 How many protons and neutrons does this atom have?

        15.2.2 What is the name of this element?

        15.2.3 Sketch the electron structure of this element using the arrow-in-circle method.

        15.2.4 What type of ion would this atom form? Give a reason for your answer.

        15.2.5 Use Lewis structures to explain how element X would react with chlorine.

    16.1 What type of bonding exists beween the atoms in the NH_3 molecule?

    16.2 Use Lewis structures to show how these bonds are formed.

    16.3 Sketch orbital diagrams to show how these bonds are formed.

17. Write balanced equations for the following chemical reactions.

    17.1 $H_2O_2 \rightarrow H_2O + O_2$

    17.2 $SO_2 + O_2 \rightarrow SO_3$

    17.3 $A\ell + O_2 \rightarrow A\ell_2O_3$

    17.4 $KC\ell O_3 \rightarrow KC\ell + O_2$

    17.5 zinc + hydrochloric acid $\rightarrow$ zinc chloride + hydrogen

    17.6 calcium hydroxide + sulphuric acid $\rightarrow$ calcium sulphate + hydrogen oxide

    17.7 nitrogen + hydrogen $\rightarrow$ ammonia

    17.8 calcium carbonate + hydrochloric acid $\rightarrow$ calcium chloride + water + carbon dioxide.

18. What are the valencies of the underlined atoms or compound ions?

    18.1 $Cu(\underline{NO_3})_2$

    18.2 $\underline{Ba}(OH)_2$

    18.3 $\underline{Na}_2O$

    18.4 $\underline{A\ell}_2O_3$

18.5 ZnS<u>O</u>₄

18.6 <u>Fe</u>Cℓ_3

18.7 <u>Mn</u>SO₄

Additional for Higher Grade

19. The electron configuration of an element may be written as $[Ar]4s^1$.

    19.1 Which element is this?

    19.2 What does [Ar] mean?

20. Draw Lewis structures for

    20.1 NH_3

    20.2 CH_4

    20.3 CO_2

    20.4 H_2O

21. If the mass of a proton is $1{,}67 \times 10^{-27}$kg, calculate the approximate mass of

    21.1 a neutron

    21.2 an alpha particle (corresponds with a Helium nucleus)

    21.3 an atom of $^{23}_{11}$Na.

    22.1 In which respect do the two isotopes of chlorine $^{35}_{17}$Cℓ and $^{37}_{17}$Cℓ differ?

    22.2 To what is their similar chemical behaviour ascribed?

    22.3 The relative atomic mass of chlorine is 35,5. Which one of the two isotopes is more abundant in nature?

    22.4 How does a chlorine atom react during a chemical reaction?

23. Why does a sugar solution not conduct an electric current while a table-salt solution does?

24. Answer the following questions on both magnesium and fluorine.

    24.1 State the number of valency electrons of each.

    24.2 Give the sp-notation for each.

    24.3 Sketch the orbitals in which the valency electrons are.

    24.4 Show, by using Lewis structures, the formation of the

        24.4.1 cation and

        24.4.2 anion
        when magnesium and fluorine bond chemically.

    24.5 What kind of force keeps the ions together in the crystal lattice?

    24.6 In which ratio is the cations and anions to be found in the crystal lattice?

    24.7 Name two properties of this kind of compound.

25.1 What do we call the modern atomic model?

25.2 What ions are formed when element number 13 reacts with bromine and what type are each of them?

25.3 Explain by means of Lewis diagrams how element number 13 would react with bromine.

25.4 What type of compound is formed in 25.3?

25.5 What do we call a reaction during which heat energy is absorbed?
Name one example of such a reaction.

ANSWERS

Section A

1. D	2. A	3. D	4. D	5. C	6. C	7. D
8. A	9. D	10. C	11. B	12. B	13. D	14. D
15. B	16. D	17. B	18. D	19. A	20. D	21. A
22. A	23. B	24. D	25. A	26. A	27. B	28. C
29. D	30. B	31. A	32. D	33. C	34. C	35. B
36. C	37. B	38. A	39. B	40. D	41. C	42. C
43. A	44. B	45. A	46. D	47. D	48. B	49. D

Section B

1.1
- An atom can only possess fixed energies described by energy levels.
- An electron can only take up or lose a certain fixed amount of energy in order to move to a higher or lower energy level.

1.2
- The model did not take into account the wave nature of electrons.
- The assumptions made in 1.1 were made without explanations for them.

2.
- There had to be a positively charged solid central particle (the nucleus).
- The mass of the particle is concentrated in this nucleus and most of the atom was empty space.

3.1 **Energy level** – the different permitted states of energy of an electron are called energy levels.
Ground state – when each electron is in the lowest possible energy level, the atom is in the ground state.

3.2
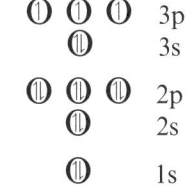

3.3 In the p sub level of the third energy level.

3.4 five

3.5 two

3.6 No. The three p orbitals in any energy level must each first have one electron before any of them may receive a second electron.

3.7 Double tearshaped, three dimensional (or two lobes forming a three dimensional figure of eight).

3.8
- a 3p orbital appears in a higher energy level than a 2p orbital.
- a 3p orbital reaches out further in space than a 2p orbital.

3.9 • an electron in a 3p orbital possesses more energy than an electron in a 3s orbital.
  • a 3s orbital is spherical while a 3p orbital has a double tear-shaped form.

4.

Atom (element)	Symbol	Atomic number	Mass number	Number of protons	Number of neutrons	Number of valence electrons	Normal valency	Ionic charge
fluorine	$^{19}_{9}F$	9	19	9	10	7	1	−1
potassium	$^{39}_{19}K$	19	39	19	20	1	1	+1
argon	$^{40}_{18}Ar$	18	40	18	22	8	0	0
magnesium	$^{24}_{12}Mg$	12	24	12	12	2	2	+2
aluminium	$^{27}_{13}A\ell$	13	27	13	14	3	3	+3
sulphur	$^{32}_{16}S$	16	32	16	16	6	2	−2
nitrogen	$^{14}_{7}N$	7	14	7	7	5	3	−3
carbon	$^{14}_{6}C$	6	14	6	8	4	4	↓

  • forms covalent bonds
  • shares electrons and does not form ions normally.

5.1 First energy level: one 1s orbital
Second energy level: one 2s orbital and three 2p orbitals

5.2 8

5.3

6.1 two

6.2 −2

6.3 2

6.4 Both are p orbitals

6.5 $X + 2e^- \rightarrow X^{2-}$

6.6.1 K_2X

6.6.2 $A\ell_2X_3$

6.7 X and K: ionic

 $A\ell$ and X: ionic

7.1 Group III

7.2 Period III

8.1 The dots or crosses represent the number of electrons in the highest energy level of the atom (i.e. the valence electrons).

8.2 Li x $\quad\quad$ $\overset{\circ}{\underset{\circ}{\text{N}}}\circ$ $\quad\quad$ $\overset{\text{x x}}{\underset{\text{x x}}{\text{x}\,\text{C}\ell\,\text{x}}}$

8.3.1 Na $\circ$ ⟵⟶ $\overset{\text{x x}}{\underset{\text{x x}}{\,\text{x}\text{F}\text{x}}}$

8.3.2 H $\circ$ ⟵⟶ $\overset{\text{x x}}{\underset{\text{x}}{\text{x}\,\text{S}\,\text{x}}}$ $\quad\to\quad$ H $\overset{\circ\;\text{x}}{\underset{\circ\;\text{x}}{\,\text{x}\,\text{S}\,\text{x}}}$

H $\circ$ ⟵ $\quad\quad\quad\quad\quad$ H

8.3.3 $\overset{\text{x x}}{\underset{\text{x x}}{\,\text{x}\text{F}\text{x}}}$⟵⟶$\overset{\circ\circ}{\underset{\circ\circ}{\circ\text{F}\circ}}$ $\quad\to\quad$ $\overset{\text{x x}}{\underset{\text{x x}}{\,\text{x}\text{F}\text{x}}}\overset{\circ\circ}{\underset{\circ\circ}{\text{F}\circ}}$

8.3.4 Mg$\overset{\text{x}}{\underset{\text{x}}{}}$⟵⟶$\overset{\circ\;\circ}{\underset{\circ\;\circ}{\circ\text{C}\ell\circ}}$

$\quad\quad\quad$⟶$\overset{\circ\;\circ}{\underset{\circ\;\circ}{\circ\text{C}\ell\circ}}$

8.3.5 $\overset{\text{x x}}{\underset{\text{x}}{\,\text{x}\text{O}\text{x}}}$⟷$\overset{\circ\circ}{\underset{\circ}{\circ\text{O}\circ}}$ $\quad\to\quad$ $\overset{\text{x x}}{\underset{\text{x}}{\,\text{x}\text{O}\,\text{x}}}\overset{\circ\circ}{\underset{\circ}{\circ\text{O}\circ}}$

8.4 $\quad$ NaF: ionic bonding
$\quad\quad$ H_2S: covalent bonding
$\quad\quad\quad$ F_2: covalent bonding
$\quad$ $MgC\ell_2$: ionic bonding
$\quad\quad\quad$ O_2: covalent bonding

9.1 Both X and Y have 10 electrons and 10 neutrons.

9.2 No. They are different elements that appear in two different groups on the Periodic Table.
R: alkali metal: Na (sodium)
X: fluorine – a halogen and non-metal.

9.3 X is an ion with 9 protons.
Y is a neutral atom with 10 protons. These two are completely different elements.

9.4 X^-: negative ionic charge, i.e. has one additional e$^-$.

9.5 X^-

9.6 Y – neon atom $\quad\quad$ S – aluminium ion.

10.1	ionic	**10.6**	ionic
10.2	covalent	**10.7**	ionic
10.3	covalent	**10.8**	ionic
10.4	ionic	**10.9**	covalent
10.5	covalent	**10.10**	ionic

11.1	$C\ell_2$	**11.6**	H_2SO_4
11.2	$Mg(OH)_2$	**11.7**	$NaNO_3$
11.3	Li_2O	**11.8**	KF
11.4	$Ca_3(PO_4)_2$	**11.9**	Na_2CO_3
11.5	$A\ell_2(SO_4)_3$	**11.10**	$Ca(NO_3)_2$

12.1	potassium sulphate	**12.5**	iron(II) oxide
12.2	calcium carbonate	**12.6**	potassium hydroxide
12.3	aluminium hydroxide	**12.7**	sodium hydrogen carbonate
12.4	zinc nitrate	**12.8**	hydrogen sulphide

13.1	non-spontaneous exothermic	**13.3**	spontaneous exothermic
13.2	spontaneous endothermic	**13.4**	spontaneous exothermic

14.1 $2H_2(g) + O_2(g) \rightarrow 2H_2O(g)$

14.2 covalent bonding

14.3 $\overset{xx}{\underset{x\,o}{x}}\overset{}{O}\overset{o}{\underset{}{x}} H$
$\quad$ H

14.4

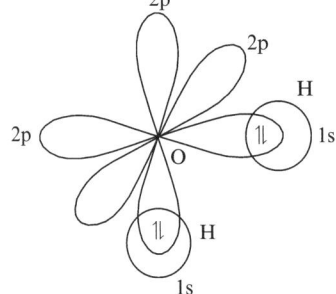

15.1.1 number of protons in the nucleus of an atom.

15.1.2 atoms of the same element having the same atomic number but different mass numbers, due to a difference in the number of neutrons.

15.1.3 a number that indicates how many bonds an atom can form when reacting with other atoms.

15.2.1 number of protons = 11
number of neutrons = 12

15.2.2 sodium

15.2.3

⑪	3s
⑪ ⑪ ⑪	2p
⑪	2s
⑪	1s

15.2.4 • positive ion or cation.
• to form a stable noble gas electron structure.

15.2.5 formation of the positive ion: $\quad$ Na$\circ$ $\quad \rightarrow \quad$ [Na]$^+$ + e$^-$

formation of the negative ion: $\quad \overset{xx}{\underset{xx}{x}}\overset{}{C}\overset{}{\underset{}{l}}\overset{x}{x} + e^- \rightarrow [\overset{xx}{\underset{xx}{x}}\overset{}{C}\overset{}{\underset{}{l}}\overset{o}{x}]^-$

packing of the ions in a crystal lattice: $\quad$ [Na]$^+$ + [$\overset{xx}{\underset{xx}{x}}Cl$$\overset{o}{x}$]$^-$ $\rightarrow$ [Na]$^+$[$\overset{xx}{\underset{xx}{x}}Cl$$\overset{o}{x}$]$^-$ $\rightarrow$ NaCl(s)

16.1 covalent bonding

16.2 $\overset{x\,x}{\underset{x}{x}}\text{N}x$ + 3H∘ → $H\overset{x\,x}{\underset{x\,o}{x}}\overset{o}{N}\overset{o}{x}H$
                           H

16.3

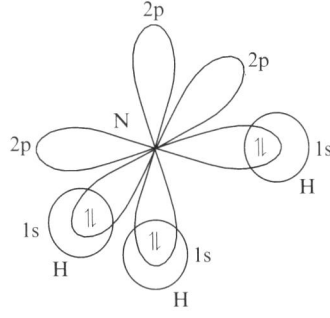

17.1 $2H_2O_2 \rightarrow 2H_2O + O_2$

17.2 $2SO_2 + O_2 \rightarrow 2SO_3$

17.3 $4A\ell + 3O_2 \rightarrow 2A\ell_2O_3$

17.4 $2KC\ell O_3 \rightarrow 2KC\ell + 3O_2$

17.5 $Zn + 2HC\ell \rightarrow ZnC\ell_2 + H_2$

17.6 $Ca(OH)_2 + H_2SO_4 \rightarrow CaSO_4 + 2H_2O$

17.7 $N_2 + 3H_2 \rightarrow 2NH_3$

17.8 $CaCO_3 + 2HC\ell \rightarrow CaC\ell_2 + H_2O + CO_2$

18.1	2		**18.5**	2
18.2	2		**18.6**	3
18.3	1		**18.7**	2
18.4	3			

Additional for Higher Grade

19.1 Potassium

19.2 Beneath the $4s^1$ valence electron there are 18 electrons which make up the electron configuration of the element argon.

20.1 $H\overset{x\,x}{\underset{x\,o}{x}}\overset{o}{N}\overset{o}{x}H$
    H

20.3 $\overset{xx}{x}\overset{}{O}\overset{x\,o}{x\,o}C\overset{o\,x}{o\,x}\overset{xx}{O}x$

20.2 $H\overset{}{\underset{x}{o}}\overset{x\,o}{C}\overset{}{\underset{o\,x}{x}}H$
     H
     H

20.4   H
    $\overset{o\,x}{o}O\overset{}{o}H$
    o o

21.1 m_n = mass of a proton = $1{,}67 \times 10^{-27}$kg

21.2 alpha particle = $^4_2\text{He}^{2+}$

$$\therefore \text{m} = 4 \times \text{mass of a proton}$$
$$= 4 \times 1{,}67 \times 10^{-27}$$
$$= 6{,}68 \times 10^{-27}\text{kg}$$

21.3 m = $23 \times$ mass of a proton   (The mass of the electrons can be ignored)
$$= 23 \times 1{,}67 \times 10^{-27}$$
$$= 3{,}84 \times 10^{-26}\text{kg}$$

22.1 $^{35}_{17}\text{C}\ell$: has 18 neutrons in the nucleus

$^{37}_{17}\text{C}\ell$: has 20 neutrons in the nucleus

22.2 the number of protons in the nucleus, namely 17

22.3 chlorine –35

22.4 forms a negative chloride ione ($\text{C}\ell^-$) by taking up an electron.

---

23. • Sugar is a covalent compound and when dissolved in water, does not conduct electricity because there are no ions to carry the electrical current.
   • Table salt is an ionic compound and when dissolved in water, forms positive and negative ions which will conduct the electricity.

24.1 The number of valency electrons: Mg – 2
$$\text{F} - 7$$

24.2 Mg: $1s^2\ 2s^2\ 2p^6\ 3s^2$
     F: $1s^2\ 2s^2\ 2p^5$

24.3

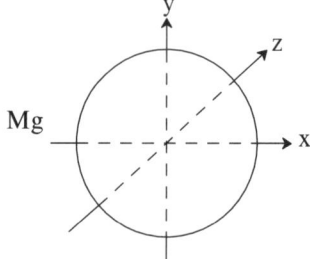
Mg

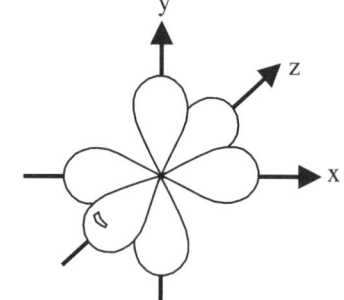
F

24.4.1 $\text{Mg:} \rightarrow [\text{Mg}]^{2+} + 2e^-$

24.4.2 $2^{x}_{x}\overset{x\,x}{\underset{x\,x}{\text{F}}}{}^{x} + 2e^- \rightarrow 2[^{x}_{x}\overset{x\,x}{\underset{x\,x}{\text{F}}}{}^{x}_{\circ}]^-$

24.5 Coulomb forces between the positive magnesium ions and the negative fluorine ions.

24.6 $\text{Mg}^{2+}: \text{F}^- = 1:2$

24.7 • high melting point

   • conducts electricity when dissolved in water or molten.

25.1 Orbital model or wave mechanical model

25.2 $_{13}Al^{3+}$: forms a positive ion or cation

$_9Br^-$: forms a negative ion or anion

25.3 Formation of a positive ion: $A\overset{x\ x}{\underset{x}{l}} \rightarrow [Al]^{3+} + 3e^-$

Formation of the negative ion: $3{\overset{\circ}{\underset{\circ\circ}{Br}}}{\overset{\circ\circ}{\circ}} + 3e^- \rightarrow 3[{\overset{\circ}{\underset{\circ\circ}{Br}}}{\overset{\circ\circ}{x}}]^-$

Ionic bonding and forming
of a crystal lattice: $[Al]^{3+} + 3[{\overset{\circ}{\underset{\circ\circ}{Br}}}{\overset{\circ\circ}{x}}]^- \rightarrow [Al]^{3+}[Br]_3^- \rightarrow AlBr_3(s)$

25.4 Ionic compound

25.5 • endothermic
• when an ionic compound dissolves in water.

9 The Reactions of the Metals and the Non-metals

Metals have always played an important role in the development of society. To recover metals from the compounds in which they occur has always been a problem. Metals are found on the left hand side of the Periodic Table.

Non-metals are found on the right hand side of the Periodic Table. Oxygen is a very active element that reacts with most metals. There are significant differences between metals and non-metals.

Metals

9.1 Reactions of metals with oxygen

The metals to be investigated are lithium (Li), sodium (Na), potassium (K), calcium (Ca), magnesium (Mg), iron (Fe) and copper (Cu).

Experiment 1: To investigate the reactions of the above mentioned metals with oxygen.

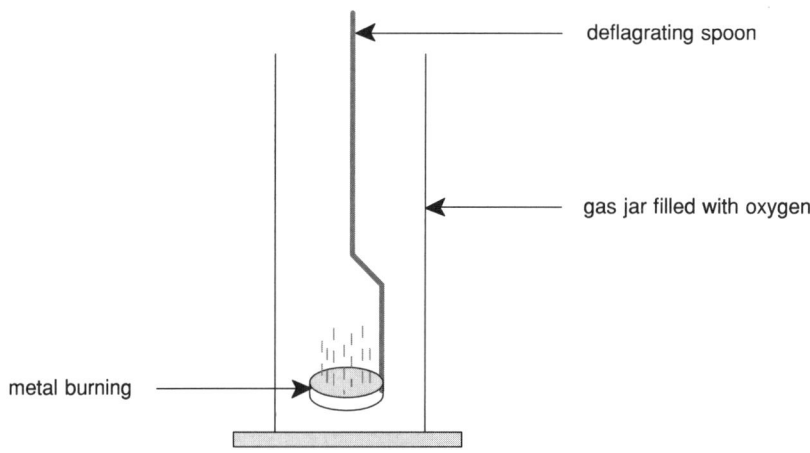

- A little piece of each of the metals is placed in a deflagrating spoon.
- Heat the spoon and its contents in the flame of a bunsen burner until it is red-hot or ignites.
- Lower the spoon into a glass jar filled with oxygen.
- Remove the spoon as soon as the reaction is completed.
- Add some purified water to the contents of the jar.
- Cover with a cover slip and shake the contents.
- Add a few drops of **bromothymol blue indicator** to the contents.

Observations and Discussion:

Element	Flame colour	Products (oxides)	Equations	Colour of the indicator	Nature of the solution
lithium	white	white powder	$4Li + O_2 \rightarrow 2Li_2O$		
		dissolves in water	$Li_2O + H_2O \rightarrow 2LiOH$	blue	alkaline
sodium	yellow	white powder	$4Na + O_2 \rightarrow 2Na_2O$		
		dissolves in water	$Na_2O + H_2O \rightarrow 2NaOH$	blue	alkaline
potassium	purple	white powder	$4K + O_2 \rightarrow 2K_2O$		
		dissolves in water	$K_2O + H_2O \rightarrow 2KOH$	blue	alkaline
calcium	brick red	white powder	$2Ca + O_2 \rightarrow 2CaO$		
		dissolves in water	$CaO + H_2O \rightarrow Ca(OH)_2$	blue	alkaline
magnesium	bright white	white powder	$2Mg + O_2 \rightarrow 2MgO$		
			$MgO + H_2O \rightarrow Mg(OH)_2$	blue	alkaline
iron	yellow glow and yellow sparks	black powder	$3Fe + 2O_2 \rightarrow Fe_3O_4$ or $4Fe + 3O_2 \rightarrow 2Fe_2O_3$		
		Insoluble in water		green	neutral
copper	red glow	black powder	$2Cu + O_2 \rightarrow 2CuO$		
		insoluble in water		green	neutral

Deductions:

- The elements which are in group I in the periodic table are very reactive (Li, Na, K)
- The elements which are in group II (Mg, Ca) in the periodic table are reactive but not as reactive as the group I elements.
- Iron and copper are least reactive.
- The indicator tests of the oxide solutions appear to be **alkaline**; they are called **basic oxides**. The high pH values (purple colour) in the case of the alkali metals show correspondence to their high degree of reactivity.

9.2 Heating of metal oxides

Another way of obtaining information about the reactivity of metals in to heat their oxides and observe how they break up into the metal and oxygen.

When **substances combine** to form chemical bands, they **release energy** to form stable compounds.

The more energy they release, the more stable the compound. To break up a compound, energy has to be put back again. The most reactive elements released the most energy to form stable oxides. They will require the most energy to break them up again.

Experiment 2: To investigate the reaction of metal oxides when heated
- Place a small amount of the following metal oxides into separate test tubes: **black copper(II) oxide** (CuO); **white magnesium oxide** (MgO); **brown lead(IV) oxide** (PbO_2) and **red mercury(II) oxide** (HgO).
- Heat each of the test tubes in a bunsen flame.
- Test for the liberation of oxygen by placing a glowing splint in the mouth of each test tube during the process of heating. (The splint ignites when oxygen is present.)

Discussion:

- **Only HgO liberated oxygen easily. Droplets of mercury** were formed against the **colder part of the test tube**. The **red colour** of the HgO changes to black. If the test tube is allowed to cool, the **colour changes back to red**

  The bonds between the mercury and the oxygen are broken and the two elements are set free:
$$2HgO \underset{\triangle}{\rightarrow} 2Hg + O_2$$

- The brown PbO_2 changes to yellow and oxygen is liberated:
$$2PbO_2 \rightarrow 2PbO + O_2$$
$$\text{brown} \triangle \text{ yellow}$$

- The CuO and MgO did not liberate oxygen but could experience a temporary change in colour.
- Now we can deduce that mercury is less reactive than magnesium and copper in relation to their reactions with oxygen.
- We can not decide on the position of lead. PbO_2 does decompose to liberate oxygen, but the oxide PbO does not decompose when heated.

At this point we can arrange the elements according to their reactivity with oxygen:

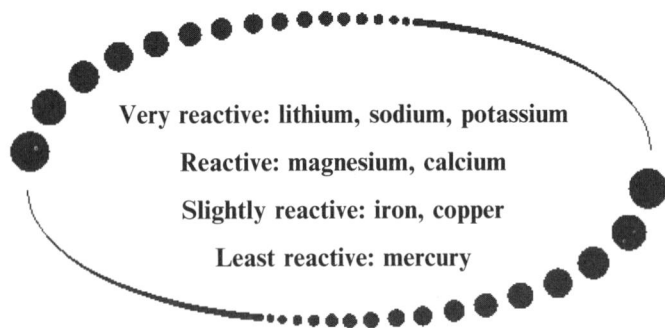

Very reactive: lithium, sodium, potassium

Reactive: magnesium, calcium

Slightly reactive: iron, copper

Least reactive: mercury

9.3 Reactions of metals with water

Experiment 3: To investigate the reactions of metals with water

A. Cold water
- Fill seven test tubes with cold water to two-thirds of their volume
- Place small pieces of the alkali metals (Li, Na, K) and small cleanly scraped platelets of Ca, Mg, Fe and Cu in turn in each test tube.
- Test for the liberation of hydrogen in each reaction by holding a burning match at the mouth of each test tube. (A slight "pop" when the gas ignites indicates the presence of hydrogen)
- Add litmus solution or litmus paper to each test tube after the reaction.

B. Hot water
- Heat the test tubes containing the metals that did not react or reacted only slightly with cold water.
- Test again for the liberation of hydrogen.
- Observe any colour changes in the indicator.

C. Steam
- Set up the apparatus as shown in the sketch.
- Use Mg, Fe and Cu powder
- Heat the water contained in the moist asbestos and the metalpowder in the test tube A simultaneously by using two bunsen burners.
- Test for the presence of hydrogen in test tube B.

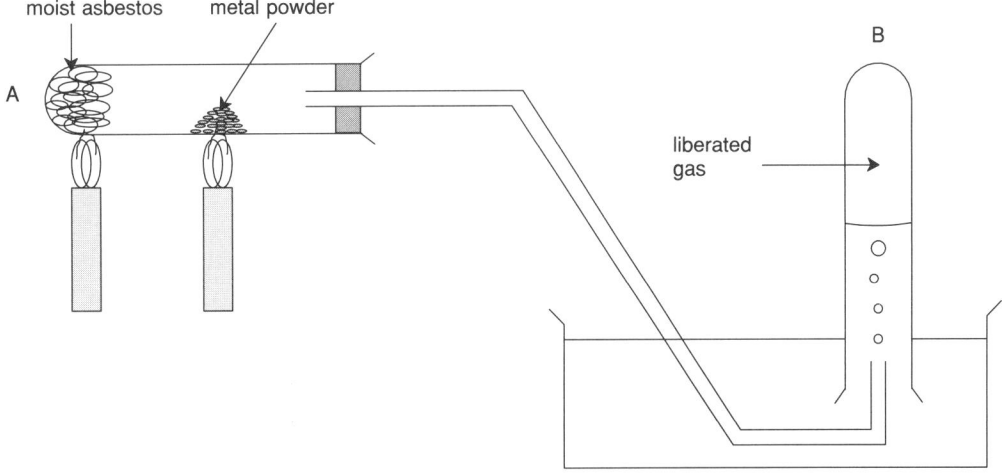

Observations and Discussion

- **The alkali metals (Li, Na, K) reacted energetically** with cold water. **Potassium ignites** and **burns with a purple flame. Sodium** moves about on the surface of the water and **melts into a little sphere. Lithium floats on the water** with a hissing sound, but **does not melt**.
  During the spontaneous reactions of the alkali metals with cold water hydrogen is liberated. The litmus solutions in the test tubes turn blue and are thus alkaline; which indicates the presence of a hydroxide.

  A comparison of the reactivities of the alkali metals indicates an increase from Li to Na to K.
  $$2Li + 2H_2O \rightarrow 2LiOH + H_2$$
  $$2Na + 2H_2O \rightarrow 2NaOH + H_2$$
  $$2K + 2H_2O \rightarrow 2KOH + H_2$$

- **Calcium** sinks to the bottom of the test tube and **bubbles are released** because of the **liberation of hydrogen gas**. The **water turns milky**; showing that the **calcium hydroxide** is **not entirely soluble in water**. The **solution turns blue** indicating that it is **alkaline**.

$$Ca + 2H_2O \rightarrow H_2 + Ca(OH)_2$$

- **Magnesium** hardly reacts with cold water. Magnesium reacts with hot water. **Hydrogen gas bubbles are liberated** and the **solution becomes slightly milky**. The litmus becomes blue which indicates that the **solution is alkaline.** When **magnesium reacts with steam a violent reaction** takes place and a **white powder is formed. Hydrogen gas is liberated.** The equations for the reactions are as follows:

Cold water: hardly no reaction ($Mg + 2H_2O \rightarrow H_2 + Mg(OH)_2$)

Hot water: $Mg + 2H_2O\,(l) \rightarrow H_2 + Mg(OH)_2$

Steam: $Mg + H_2O\,(g) \rightarrow H_2 + MgO$.

- **Iron** shows no visible reaction with cold or hot water. It glows brightly in steam. **Hydrogen gas is liberated** and a **black powder** (magnetic iron oxide) **is formed**.

$$3Fe + 4H_2O(g) \rightarrow Fe_3O_4 + 4H_2$$

- **Copper shows no reaction with cold water, hot water or steam**

All the metals, except copper, were able to displace hydrogen from hydrogen oxide (water). These metals are considered to be more reactive than hydrogen. The reactivity series which can now be compiled appears to be:

K Na Li Ca Mg Fe (H) Cu Hg

←————————————————————————

most reactive   least reactive

9.4 The reaction of metals with solutions of metal salts

When we mix metal A with a solution of the salt of metal B and a reaction takes place during which metal A takes the place of metal B, we can say that A is more reactive than B, because its atoms form stronger bonds than those of B.

A more reactive element will displace a less reactive element from a solution of its compound.

Experiment 4: To investigate the displacement of one metal from a solution of its salt by another metal.

- The reactions of iron (Fe), zinc (Zn), magnesium (Mg) and copper (Cu) with solutions of iron(II) sulphate, zinc sulphate, magnesium sulphate and copper sulphate are going to be investigated.
- Add approximately 20 cm^3 of each of the solutions to test tubes.
- Add a little of the powder of each metal to each solution in turn.
- Take the temperature of each test tube.
- Record all observations and tabulate the results. A tick indicates that a reaction takes place. A cross indicates that no reaction takes place.

Results:

Metal	Solutions of metal salts			
	FeSO$_4$	ZnSO$_4$	MgSO$_4$	CuSO$_4$
Iron Fe		x	x	✓ Blue becomes colourless Temperature increase Red copper deposit
zinc Zn	✓ Green becomes colourless. Temperature increase		x	✓ Blue becomes colourless Temperature increase Red copper deposit
magnesium Mg	✓ Green becomes colourless. Temperature increase	✓ Temperature increase		✓ Blue becomes colourless Temperature increase Red copper deposit
copper Cu	x	x	x	

Deductions:

• The more substances a metal react with, the more reactive it is. .

• The order of reactivity is   Mg   Zn   Fe   Cu

    ⟵

    most reactive   least reactive

• The chemical equations for these reactions are:
    Fe + CuSO$_4$ → FeSO$_4$ + Cu
    Zn + FeSO$_4$ → ZnSO$_4$ + Fe
    Zn + CuSO$_4$ → ZnSO$_4$ + Cu
    Mg + FeSO$_4$ → MgSO$_4$ + Fe
    Mg + ZnSO$_4$ → MgSO$_4$ + Zn
    Mg + CuSO$_4$ → MgSO$_4$ + Cu

• During these reactions the temperature of the reagents rises.
    This indicates that **energy is liberated**. These **reactions are exothermic**.

• These reactions are redox reactions (see Chapter 11).

From all the experiments that have been done the following **reactivity series for metals and hydrogen can be deduced**.

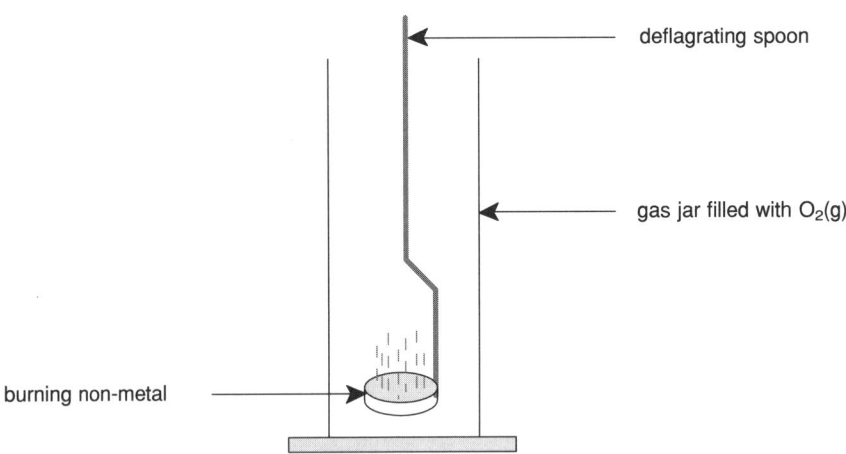

potassium
sodium
lithium
calcium
magnesium
zinc
iron
lead
(hydrogen)
copper
mercury

most reactive / least reactive

From Mg downwards any metal displaces less reactive metals from solutions of their salts.

- The most reactive metals are in Group I
- Reacitivity of metals decreases from left to right in a period
- Metals higher up in the same group are less reactive than those lower down. (Reactivity increases from top to bottom in a group).
- As being indicated alongside the reactivity series above, the rule applies that any metal displaces a less reactive metal **lower down** from a solution of the latter's salt, e.g. magnesium displaces zinc from a solution of zinc chloride or iron displaces mercury from a solution of mercury nitrate, etc.
- Above magnesium in the activity series, the metals are so reactive that they will react with the water and not with the salt solution.

Non-metals
9.5 Reaction of non-metals with oxygen.
Experiment 5: To investigate the reactions of non-metals with oxygen.

deflagrating spoon

gas jar filled with $O_2(g)$

burning non-metal

- Place a small amount of charcoal powder, flowers of sulphur and red phosphorus in a deflagrating spoon in turn.
- Heat the spoon and its contents in the flame of a bunsen burner until it ignites.
- Lower the spoon into a gas jar filled with oxygen and which also contains a small amount of water.
- Remove the spoon as soon as the reaction is completed.
- Cover with a cover slip and shake the jar so that the gas can dissolve in the water.
- Add a few drops of bromothymol blue indicator to the contents of the jar.

Observations and discussion

Element	Flame colour	Products formed	Equations of reactions	Colour of indicator	Nature of the solution	Name of the product
carbon	yellow	colourless gas	$C + O_2 \rightarrow CO_2$			carbon dioxide
		gas dissolves in water	$CO_2 + H_2O \rightarrow H_2CO_3$	yellow	acidic	carbonic acid
sulphur	blue	colourless gas	$S + O_2 \rightarrow SO_2$			sulphur dioxide
		gas dissolves in water	$SO_2 + H_2O \rightarrow H_2SO_3$	yellow	acidic	sulphurous acid
phosphorous	yellow	dense white fumes	$P_4 + 5O_2 \rightarrow 2P_2O_5$			phosphorus pentoxide
		dissolves in water	$P_2O_5 + 3H_2O \rightarrow 2H_3PO_4$	yellow	acidic	phosphoric acid

- When bromothymol blue is placed in a solution of a non-metal oxide it turns yellow. **Non-metal oxides** produce **acidic solutions** and are called **acidic oxides**.
- **Metal oxides** are called **basic oxides.**

9.6 Reactions of the halogens with metals

Die term **halogens** refers to the elements in group VII of the Periodic Table. When we study the physical properties of these elements at room temperature, the following are observed:
$C\ell_2$ **(chlorine)** is a poisonous yellow-green gas.
Br_2 **(bromine)** is a volatile reddish brown liquid.
F_2 **(fluorine)** is a pale yellow gas
I_2 **(iodine)** is a solid occuring as flaky crystals with a black metallic glimmer.
The halogens all occur as **diatomic molecules**. The **high degree of reactivity** of these elements gives rise to their **toxic** nature.

Experiment 6: To investigate the reaction of iodine with some metals.

- Crush iodine crystals to a fine powder.
- Place a small piece of sodium, a small amount of aluminium powder, a small amount of magnesium powder, and a small amount of zinc powder on each of four small tiles.
- Cover each of the Mg, Al, Zn and Na with iodine powder.
- Drip a few drops of water onto the iodine.

Results and Discussion
- The **sodium** ignites, **purple fumes are liberated** and a **white powder, sodium iodide**, is formed:
  $2Na + I_2 \rightarrow 2NaI$
- **Aluminium** and **magnesium react less vigorously** than the sodium. Aluminium is less reactive than the magnesium. **Purple iodine fumes** form:
  $Mg + I_2 \rightarrow MgI_2$
  $2A\ell + 3I_2 \rightarrow 2A\ell I_3$
- **Zinc** also reacts with the iodine but it takes longer to start. Zn is less reactive than aluminium. **Purple iodine fumes** also appear after a while:
  $Zn + I_2 \rightarrow ZnI_2$

- The reactivities of these four metals with iodine correspond with the reactivity series for metals (section 9.4)
- The **water acts as a catalyst**.
- The purple iodine fumes indicate that **energy is liberated** and that these reactions are **exothermic**.

9.7 Reactions of the halogens with solutions of their salts

We shall **compare the reactivity of three halogens**, i.e. **chlorine, bromine** and **iodine**. Fluorine will not be investigated as it is so reactive that it combines with most elements. Fluorine reacts very vigorously with water.

Experiment 7: To test for the presence of iodine and bromine.

- Halogens are identified by the characteristic coloured solutions formed with carbon disulphide (CS_2), chloroform or xylene.
- Pour small quantities of iodine, bromine water and chlorine water in each of three test tubes. Shake it up with about $5cm^3$ of carbon disulphide, chloroform or xylene.

Results:

- There forms two layers in each of the three test tubes. A water layer on top and a carbon disulphide or chloroform layer at the bottom. CS_2 and chloroform have greater densities than water. (Xylene is less dense than water and floats on top of the water)
- Bromine – the bottom layer gets a **yellow-brown** colour
- Iodine – the bottom layer becomes **purple**
- Chlorine – the bottom layer **remains clear**.

Experiment 8.1: To investigate the reactions of chlorine with halide solutions.

- Pour 3 cm^3 of solutions of potassium chloride, potassium bromide and potassium iodide in three different test tubes.
- Add 2 cm^3 xylene or chloroform to each of the three test tubes.
- Add chlorine water to each test tube and shake.

Results:

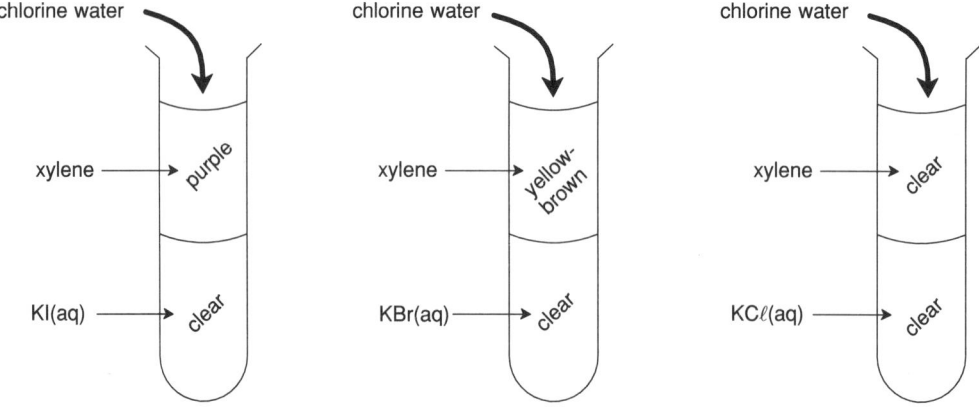

Experiment 8.2: To investigate the reactions of bromine with halide solutions.

- Pour 3 cm^3 of solutions of potassium chloride, potassium bromide and potassium iodide in three different test tubes.
- Add 2 cm^3 of xylene to each of the three test tubes.
- Add clear solutions of very dilute bromine water to each test tube and shake.

Results:

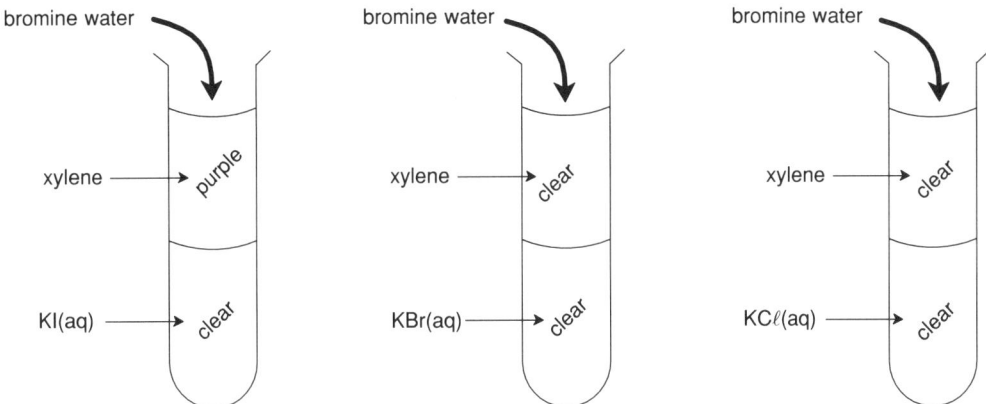

- **NB:** If chloroform or carbon disulphide are used the water layer will be on top.
- Cl_2 displaces both I_2 and Br_2 from their potassium halide compounds.
- Br_2 displaces only I_2 from potassium iodide
- Chlorine is the most reactive, followed by bromine and then iodine.
- The reactivities of the halogens decrease from top to bottom in the Periodic Table.

The results are summarised in the table that follows:

Element (Halogen)	Halide solution			Equations	Colour of halogen in chloroform
	KCℓ	KBr	KI		
Cl_2	x	✓	✓	$2KBr + Cl_2 \rightarrow 2KCl + Br_2$	yellow brown
				$2KI + Cl_2 \rightarrow 2KCl + I_2$	purple
Br_2	x	x	✓	$2KI + Br_2 \rightarrow 2KBr + I_2$	purple

The equations of the three reactions in the table can easily be deduced from the Table of Redox Half Reactions in Chapter 11 and are therefore not necessary to be memorised.

The Periodic Table

- A halogen can displace another halogen below it in the Periodic Table from a solution of its compound.
- For non-metals, reactivity decreases from top to bottom in a group and increases from left to right across a period.

QUESTIONS

Section A

Various possibilities are suggested as answers to the following questions. Indicate the correct answer.

1. Sodium, potassium and lithium are known as . . .
   A alkalis.
   B indicators.
   C acids.
   D alkali metals.

2. Which of the following metals will react most spontaneously with oxygen when heated?
   A K
   B Na
   C Ca
   D Mg

3. Which of the following oxides will change a red litmus solution blue?
   A CaO
   B NO
   C SO_2
   D CO_2

4. Which of the following metals will react with cold water to form hydrogen?
   A Silver
   B Lithium
   C Mercury
   D Aluminium

5. Which of the following metals does not react with cold water?
   A Potassium
   B Lithium
   C Iron
   D Magnesium

6. The products formed when potassium reacts with water, are . . .
   A $K_2O + H_2$.
   B $KOH + H_2$.
   C $KO_2 + H_2$.
   D $KOH + O_2$.

7. When metals react with water, they form . . .
   A hydroxides which have no reaction on litmus.
   B oxides which colour blue litmus red.
   C hydroxides which colour red litmus blue.
   D hydroxides which colour blue litmus red.

8. A powder of which of the following metals is used in flares because it produces a brilliant white light?
   A Silver
   B Sodium
   C Potassium
   D Magnesium

9. The following metals are arranged according to their increasing chemical reactivity.

   Mg   Ca   Zn   Fe   Pb

   $\longleftarrow$

   most reactive   least reactive

   Which of the following reactions will occur spontaneously?

   A $Pb + FeSO_4 \rightarrow PbSO_4 + Fe$
   B $Zn + CaSO_4 \rightarrow ZnSO_4 + Ca$
   C $Mg + ZnSO_4 \rightarrow MgSO_4 + Zn$
   D $Ca + MgSO_4 \rightarrow CaSO_4 + Ca$

10. Which of the following chemical substances will yield oxygen when heated?
    A MgO
    B HgO
    C ZnO
    D Na_2O

11. An unknown element is given to you. It has the following properties:
    I. It reacts with oxygen when heated
    II. It reacts vigorously with water to form a hydroxide.

    The element is . . .

    A lithium.   C nitrogen.
    B potassium.   D magnesium.

12. Which of the following oxides will change the colour of blue litmus to red?
    A MgO   C Na₂O
    B K₂O   D SO₂

13. Sulphurous acid (H₂SO₃) is formed during the reaction between . . .
    A sulphur(IV) oxide and water.   C sulphur and water.
    B sulphur and oxygen.   D sulphur(IV) oxide and hydrogen.

14. When non-metals burn in oxygen, they form . . .
    A hydroxides which dissolve in water to form acids.
    B hydroxides which dissolve in water to form alkalis.
    C oxides which dissolve in water to form hydroxides.
    D oxides which dissolve in water to form acids.

15. Phosphorus is stored under water because it . . .
    A will decompose in bright light.
    B will react vigorously with the oxygen in the air.
    C liberates phosphorus(V) oxide.
    D shows acidic properties.

16. Carbon dioxide reacts with water and forms . . .
    A carbon.   C carbon monoxide.
    B carbonic acid.   D carbon dioxide.

17. Another name for an acidic oxide is . . .
    A acid anhydride.   C oxidizing agent.
    B alkali.   D halide.

18. Which of the following statements is true for iodine?
    A Iodine melts when heated.
    B Iodine cannot displace chlorine from a chloride solution.
    C Iodine dissolved in CS₂ has a brown colour.
    D Iodine is a diatomic gas.

19. Which of the following reactions is incorrect?
    A $2K + I_2 \rightarrow 2KI$ C $2A\ell + 3I_2 \rightarrow 2A\ell I_3$
    B $2KC\ell + Br_2 \rightarrow 2KBr + C\ell_2$ D $2KI + Br_2 \rightarrow 2KBr + I_2$

20. The colour of chlorine, bromine and iodine respectively when dissolved in chloroform is . . .
    A red, purple and yellow.   C colourless, yellow-brown and purple.
    B colourless, purple and   D yellow-brown, purple and colourless.
       yellow-brown.

Section B

1. Potassium is burnt in oxygen.
    1.1 Write down a balanced chemical equation for the reaction and describe what is observed during the reaction.

165

1.2 The product of the reaction is dissolved in water and a few drops of bromothymol blue indicator are added. Describe what would be observed.

1.3 Write down a balanced chemical equation to explain your answer in 1.2.

2. **2.1** What are "basic oxides"?
    2.2 How will you prove that calcium (Ca) forms a basic oxide? Use chemical equations to prove your statements.

3. Complete and balance the following chemical equations.
    3.1 $Li + O_2 \rightarrow$
    3.2 $Cu + O_2 \rightarrow$
    3.3 $Na_2O + H_2O \rightarrow$
    3.4 $Fe_2O_3 + H_2O \rightarrow$
    3.5 $CuO + H_2O \rightarrow$

4. **4.1** Sodium, potassium and lithium are stored in a certain liquid. Why is this liquid used for this purpose? Give the name of this liquid.
    4.2 What will you observe when you cut a piece of potassium? Explain the nature of both the inside and the outside of the piece of potassium.
    4.3 Which elements can be classified as alkaline earth metals?
    4.4 Sodium is not found in its pure state in nature but silver is. Explain why.

5. **5.1** Give a test for oxygen.
    5.2 Mercury(II) oxide and magnesium oxide are heated in a test tube. Write chemical equations for the reactions that take place and also indicate all the phases.
    5.3 Does the mercury(II) oxide undergo a permanent or a temporary change? Explain.
    5.4 Will the mass of the test tube containing the mercury(II) oxide be more or less after being heated than before? Explain your answer.
    5.5 Why should one perform the experiment with the mercury(II) oxide in a fume chamber?
    5.6 Which metal, mercury or magnesium is more reactive? Give a reason for your answer.
    5.7 Is mercury(II) oxide a mixture or a compound of mercury and oxygen? Give three possible reasons.

6. A piece of magnesium is placed in a beaker containing water.
    6.1 What will you observe?
    6.2 The beaker with its contents is now heated. What will you observe now?
    6.3 A violent reaction takes place between magnesium and steam. State all observations that can be made.
    6.4 Write balanced equations for the following reactions:
        6.4.1 magnesium and warm water
        6.4.2 magnesium and steam.

7. Equal sized pieces of lithium, potassium and copper are placed in separate test tubes containing cold water. A few drops of bromothymol blue are added to each test tube.
    7.1 In which cases do reactions occur?
    7.2 Give balanced equations for the chemical reactions which occur.
    7.3 Which was the fastest reaction?
    7.4 Give the colour of the solutions in the test tubes? Explain.
    7.5 Arrange these metals in order of increasing chemical reactivity.
    7.6 Which gas is liberated during the reactions?
    7.7 How do you test for this gas?

8. A small piece of aluminium is put into a solution of copper(II) sulphate.
    8.1 What will you observe?

8.2 Write down a balanced chemical equation for the reaction taking place.

9. The imaginary elements A, B, C and D are added to solutions of the compounds ASO_4, BSO_4, CSO_4 and DSO_4 respectively. The reactions which took place are marked with a x in the table.

	A	B	C	D
ASO_4		X	X	
BSO_4				
CSO_4		X		
DSO_4	X	X	X	

9.1 Arrange the elements in order of increasing chemical reactivity.

9.2 Complete the following chemical equations from the information in the table:
B + ASO_4 →
D + CSO_4 →
C + DSO_4 →

10. A piece of yellow phosphorous is heated over a flame and lowered into a cylinder containing oxygen.
10.1 In which liquid is phosphorous kept in the laboratory? Explain why.
10.2 How would you cut a small piece of phosphorous from a bigger piece?
10.3 Is it necessary to heat phosphorous before immersing it into a gas cylinder containing oxygen? Give a reason for your answer.
10.4 What will you observe when the phosphorous reacts with the oxygen?
10.5 Give the name of the product formed.
10.6 Write a balanced chemical equation for the reaction taking place?
10.7 The product formed in 10.4 is then dissolved in water. Give the chemical equation of the reaction taking place.
10.8 Litmus paper is added to the solution in 10.7. What is the colour of the litmus paper?
10.9 What is your conclusion about non-metal oxides?

11. Complete the following chemical equations and give the names of the products:
11.1 $S + O_2$ →
11.2 $C + O_2$ →
11.3 $SO_2 + H_2O$ →
11.4 $CO_2 + H_2O$ →

12. Iodine and magnesium both in powder form, are mixed but no reaction takes place.
12.1 What would you add to the mixture to start the reaction?
12.2 Why is it necessary to add this substance?
12.3 Is this reaction endo- or exothermic? Give a reason for your answer.
12.4 What will you observe when the reaction takes place?
12.5 Write a balanced equation for the reaction between iodine and magnesium.

13. A concentrated solution of chlorine water is added to solutions of KI and KBr respectively.
13.1 Explain everything that you would observe.
13.2 Write balanced equations for both reactions.

14.   **14.1** What is the name of the elements of group VII?

    14.2 Complete the following chemical equations where a reaction does occur:

$KC\ell\,(aq) + Br_2 \rightarrow$

$KI\,(aq) + Br_2 \rightarrow$

$KC\ell\,(aq) + I_2 \rightarrow$

$KBr\,(aq) + I_2 \rightarrow$

    14.3 Write down the halogens in order of increasing chemical reactivity.

15.   On the shelf of the laboratory are two bottles with damaged labels. The substances are halides of potassium. Chlorine water and CS_2 are added to solutions of both bottles marked A and B. The CS_2 in A turns yellowish-brown and that in B purple.

    15.1 Name the halides in both A and B.

    15.2 Write balanced equations for the reactions.

    15.3 Which one of the CS_2 or the solution would form the bottom layer in the test tube? Give a reason for your answer.

ANSWERS

Section A

1. D	2. A	3. A	4. B	5. C	6. B	7. C
8. D	9. C	10. B	11. B	12. D	13. A	14. D
15. B	16. B	17. A	18. B	19. B	20. C	

Section B

1.1 $4 K + O_2 \rightarrow 2K_2O$
Potassium burns with a purple flame and forms a white powder.

1.2 The bromothymol blue indicator is blue in the solution. This indicates that the solution is alkaline because KOH is formed.

1.3 $K_2O + H_2O \rightarrow 2KOH$

2.1 Oxides of metals react with water to form alkalis

2.2
- Place a piece of calcium in a deflagrating spoon
- Heat in a flame until it ignites (reacts with oxygen).
- Put it into a gas jar containing oxygen
  $2Ca + O_2 \rightarrow 2CaO$
- Add a few cm^3 water to the mixture in the gas jar and shake.
  $CaO + H_2O \rightarrow Ca(OH)_2$
- Test the solution with a piece of litmus paper. The litmus will turn blue because of the alkali formed.
- Calcium thus forms a basic oxide (CaO)

3.1 $4Li + O_2 \rightarrow 2Li_2O$
3.2 $2Cu + O_2 \rightarrow 2CuO$
3.3 $Na_2O + H_2O \rightarrow 2NaOH$
3.4 $Fe_2O_3 + 3H_2O \rightarrow 2Fe(OH)_3$
3.5 $CuO + H_2O \rightarrow$ no reaction (CuO is insoluble in water).

4.1 The alkali metals react vigorously with water and oxygen in the air. It must therefore be stored in a liquid that does not contain oxygen or water. The liquid is called paraffin.
4.2 The freshly cut surface appears shiny, but after a while it becomes matt. The potassium reacts with the oxygen in the air and potassium oxide forms.
4.3 Group 2 elements e.g. magnesium and calcium.
4.4 Sodium is chemically very reactive. It would react with oxygen and water in both the air and ground to form sodium oxide and sodium hydroxide respectively. Silver is chemically very unreactive.

5.1 A glowing splint will ignite in oxygen.

5.2 $2HgO(s) \xrightarrow{\triangle} 2Hg(\ell) + O_2(g)$

$MgO(s) \xrightarrow{\triangle}$ no permanent change or reaction.

5.3 **Temporary:** the colour changes from red to black to red again and
permanent change: new substances with new chemical properties are formed

5.4 **Less:** The oxygen formed escaped from the test tube.

5.5 Mercury(II) oxide and mercury vapour are both very poisonous.

5.6 Magnesium is more reactive than mercury and holds onto its oxygen more strongly. Mercury releases its oxygen quite readily and seems to be less reactive.

5.7 **Compound:** • a chemical reaction is necessary to prepare mercury(II) oxide.

         • it can only be broken down into mercury and oxygen by chemical methods.

         • it has properties that differ from those of the mercury and oxygen.

6.1 The magnesium hardly reacts with the cold water. A few bubbles of hydrogen gas is formed.

6.2 Bubbles of hydrogen gas is formed. The solution becomes milky.

6.3 The magnesium glows brightly. A white powder is formed. A gas is formed. (When tested with a burning match it proves to be $H_2(g)$).

    6.4.1 $Mg + 2H_2O(\ell) \rightarrow Mg(OH)_2 + H_2(g)$

    6.4.2 $Mg + H_2O(g) \rightarrow MgO(s) + H_2(g)$

7.1 Only Li and K react with the cold water.

7.2 $2Li + 2H_2O \rightarrow 2LiOH + H_2$

    $2K + 2H_2O \rightarrow 2KOH + H_2$

7.3 Potassium with water

7.4 Blue. LiOH and KOH are alkaline

7.5 Cu, Li, K

7.6 Hydrogen (H_2)

7.7 Hold a burning match near the mouth of the test tube. A cracking sound proves the presence of hydrogen.

8.1 The blue colour of the $CuSO_4$-solution disappears and a brown layer is formed on the $A\ell$.

8.2 $2A\ell + 3CuSO_4 \rightarrow A\ell_2(SO_4)_3 + 3Cu$

9.1 D, A, C, B.

9.2 $B + ASO_4 \rightarrow BSO_4 + A$

    $D + CSO_4 -$ no reaction

    $C + DSO_4 \rightarrow CSO_4 + D$

10.1 Water

    Phosphorus reacts spontaneously with the oxygen in the air.

10.2 Cut off a small piece under the water and remove it by using a pair of tongs

10.3 No. It reacts spontaneously with oxygen

10.4 • Burns with a clear yellow flame in oxygen

    • Dense white fumes are formed

10.5 Phosphorus(V) oxide (P_2O_5)

10.6 $P_4 + 5O_2 \rightarrow 2P_2O_5$

10.7 $P_2O_5 + 3H_2O \rightarrow 2H_3PO_4$

10.8 Red

10.9 It performs like acids (acidic oxides).

11.1 $S + O_2 \rightarrow SO_2$ sulphur dioxide or sulphur(IV) oxide

11.2 $C + O_2 \rightarrow CO_2$ carbon dioxide or carbon(IV) oxide

11.3 $SO_2 + H_2O \rightarrow H_2SO_3$ sulphurous acid

11.4 $CO_2 + H_2O \rightarrow H_2CO_3$ carbonic acid

12.1 water

12.2 water is the catalyst needed to start the reaction.

12.3 exothermic/energy is liberated in the form of heat.

12.4 purple iodine fumes forms – this indicates that the temperature of the mixture increases.

12.5 $Mg + I_2 \rightarrow MgI_2$

13.1 The colourless KI solution will turn light brown because I_2 has formed
The colourless KBr solution will turn brown because Br_2 has formed.

13.2 $2KI + C\ell_2 \rightarrow 2KC\ell + I_2$
$2KBr + C\ell_2 \rightarrow 2KC\ell + Br_2$

14.1 halogens

14.2 $KC\ell + Br_2$ – no reaction
$2KI + Br_2 \rightarrow 2KBr + I_2$
$KC\ell + I_2$ – no reaction
$KBr + I_2$ – no reaction

14.3 Iodine, bromine, chlorine

15.1 A: KBr
B: KI

15.2 A: $2KBr + C\ell_2 \rightarrow 2KC\ell + Br_2$
B: $2KI + C\ell_2 \rightarrow 2KC\ell + I_2$

15.3 CS_2
CS_2 is more dense than water.

10 *Acids, Bases and Salts*

This chapter is concerned with ways to recognise groups of compounds with the same characteristics and distinguish between them. In Grades 7 and 9 you have already identified acids, bases and salts. An empirical approach will be used to look more closely into their characteristics like general properties, effect on indicators and reactions with other substances. This study will prepare learners to study acids and bases by using theoretical models in Grade 12.

1. Characteristics of acids

1.1 Naming acids

Name of compound	Formula	General name	Strength
Hydrogen chloride	$HC\ell$	Hydrochloric acid	Strong
Hydrogen sulphate	H_2SO_4	Sulphuric acid	Strong
Hydrogen nitrate	HNO_3	Nitric acid	Strong
Ethanoic acid	CH_3COOH	Vinegar	Weak

Table 1

- All acids contain hydrogen. The presence of hydrogen is responsible for the characteristic properties of acids.
- When acids are diluted in water, the temperature rises. This indicates that a reaction is taking place.
- All acids diluted in water form H_3O^+ ions.

$$H^+ \;+\; H_2O \;\rightleftharpoons\; H_3O^+$$

- Some acids are completely dissociated when diluted. They are called strong acids.
- Weak acids only dissociate partially.

"Strong" relates to the degree of dissociation and not to the concentration of H_3O^+ or pH of an acid.

NB: It is important to note that in Grade 10 we are working with dilute acids only, i.e. acids that are diluted with water to some extent.

1.2 Taste

- Dilute acids have a sour taste, like lemon juice or vinegar. This can be demonstrated by tasting vinegar and tartaric acid. Do not taste the mineral acids that you find in the laboratory.

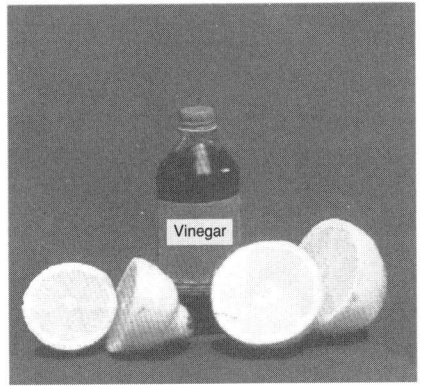

Fig. 1   Some acids for household use

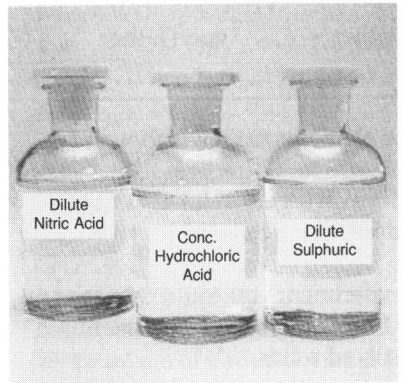

Fig. 2   Bottles with common laboratory acids

1.3 Effect on indicators

- Acids change the colour of indicators. If an indicator is added to an acid, the following colour changes are observed:

Indicator	Colour in neutral solution	Colour in acid
Litmus	Purple	Red
Bromothymol blue	Green	Yellow
Phenolphthalein	Rose pink	Colourless
Methyl orange	Orange-red	Red
Universal indicator	Green	Red

Table 2

Solution acidic   Neutral solution   Solution basic

| 1 | 2 | 3 | 4 | 5 | 6 | 7 | 8 | 9 | 10 | 11 | 12 | 13 | 14 |

Fig. 3   Universal pH-scale

1.4 Conducting electricity

All acids diluted in water form H_3O^+ ions. These ions are charge carriers in solutions.
- Aqueous solutions of acids conduct electricity.

1.5 pH-scale

The pH of a solution indicates the degree of acidity or alkalinity of a solution. It also indicates the concentration of H_3O^+ ions.
- The pH of an acidic solution is smaller than 7.
- As the concentration of H_3O^+ ions of an acid increases, its pH decreases.

1.6 Reaction of acids with

1.6.1 Metals

During investigations of the reactions between dilute acids, zinc granules and magnesium ribbon the following are observed:

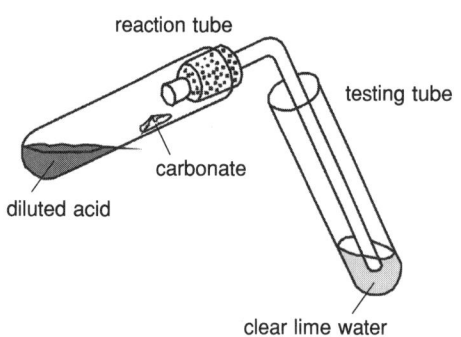

reaction tube

testing tube

carbonate

diluted acid

clear lime water

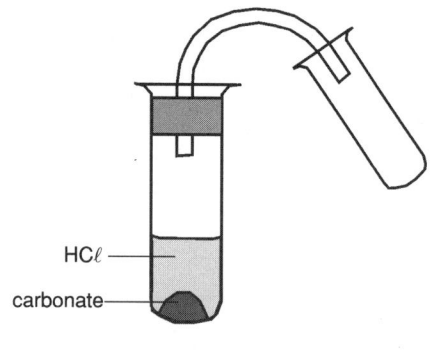

HCℓ

carbonate

Fig. 4 How to test for the gas that is liberated during the reaction of acids and carbonates

Fig. 5 A method to collect the gas that is liberated

- Acids react with certain metals to liberate a gas.
- Test the gas with a burning match; an explosive sound is heard.
- The gas is hydrogen.
- Acids produce hydrogen.
- Not all metals can displace hydrogen from acids; only those that are more reactive than hydrogen
- Metal + acid → salt + hydrogen.

Examples of reactions:

$$Mg(s) + 2HC\ell \rightarrow MgC\ell_2 + H_2(g)$$
$$Zn(s) + 2HNO_3 \rightarrow Zn(NO_3)_2 + H_2(g)$$
$$Ag(s) + H_2SO_4 - \text{no reaction}$$

Balancing the equations of reactions:

Zinc granules react with dilute nitric acid. Write a balanced equation for the reaction.

1. Write down the formulae of the reactants: $Zn + HNO_3 \rightarrow$

2. Identify the reactants: metal + acid

3. Use general reaction table and add the products: metal + acid → salt + hydrogen
   $Zn + HNO_3 \rightarrow$ $+ H_2$

4. Identify ions to form salt: Zn^{2+} and NO_3^-

5. Make sure product is neutral: (use table of compound ions) $Zn(NO_3)_2$

6. Count the number of atoms and balance: $Zn + 2HNO_3 \rightarrow Zn(NO_3)_2 + H_2$

1.6.2 Carbonates

During investigation of the reactions between dilute acids and sodium carbonate, the following are observed:

174

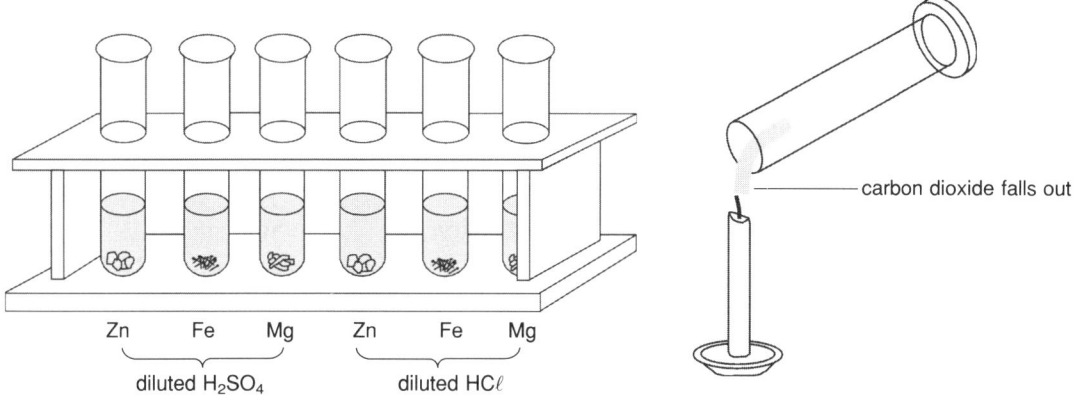

Fig 6.1   Reactions of dilute acids with metals   Fig. 6.1   One can pour carbon dioxide just like a liquid

- Acids liberate a gas from carbonates.
- Test the gas with clear lime water; it turns milky.
- The gas is carbon dioxide.
- The acidic properties are destroyed.
- Carbonate + acid → salt + water + carbon dioxide

Examples of reactions:

$$Na_2CO_3 + 2HC\ell \rightarrow 2NaC\ell + H_2O + CO_2(g)$$
$$Na_2CO_3 + 2HNO_3 \rightarrow 2NaNO_3 + H_2O + CO_2(g)$$

1.6.3 Metal oxides

During the investigation of the reactions between dilute acids and metal oxides the following are observed:

- Acids react with metal oxides to produce water.
- Test the product with blue cobalt chloride paper; it turns pink.
- The solid metal oxide dissolves to form a solution with a different colour.
- The acidic properties are destroyed if an excess of metal oxide is added.
- Metal oxide + acid → salt + water.

Examples of reactions:

$$CuO(s) + H_2SO_4 \rightarrow CuSO_4(aq) + H_2O$$
(black)   (blue)
$$MgO(s) + 2HC\ell \rightarrow MgC\ell_2(aq) + H_2O$$
(white)   (colourless)

1.6.4 Metal hydroxides

During the investigation of the reactions between dilute acids and metal hydroxides the following are observed:

- Acids react with metal hydroxides to liberate energy. The temperature of the mixture rises.
- The acidic properties are destroyed when acids are added to metal hydroxides.
- Metal hydroxide + acid → salt + water

Examples of reactions:

$$2KOH + H_2SO_4 \rightarrow K_2SO_4(aq) + 2H_2O$$
$$Mg(OH)_2 + 2HC\ell \rightarrow MgC\ell_2(aq) + 2H_2O$$

2. Characteristics of bases

2.1 Naming bases:

Name of compound	Formula	Solubility	Strength
Sodium hydroxide	NaOH	Soluble	Strong
Ammonia	NH_3	Soluble	Weak
Magnesium hydroxide	$Mg(OH)_2$	Soluble	Strong
Potassium carbonate	K_2CO_3	Soluble	Weak
Magnesium oxide	MgO	Less soluble	Weak
Copper(II) oxide	CuO	Insoluble	Weak

Table 3

- Bases that are soluble in water are called alkalis.
- When bases are diluted with water, the temperature rises. This indicates that a reaction is taking place.
- The hydroxide ions, OH^-, that are liberated are responsible for the characteristic properties of alkalis.

$$\text{(dissociates in water)}$$
$$NaOH(s) \longrightarrow Na^+(aq) + OH^-(aq)$$
$$\text{(reacts with water)}$$
$$NH_3(g) \longrightarrow NH_4^+(aq) + OH^-(aq)$$

- Some alkalis are completely ionised when diluted. They are called strong bases.
- Weak bases ionize only partially.

"Strong" relates to the degree of dissociation and not to the concentration of OH^+ ions or pH of a base

2.2 Feeling:

- Dilute bases feel soapy when rubbed between your fingers.

2.3 Effect on indicators:

- Alkalis change the colour of indicators. When indicators are added to alkalis, the following colour changes are observed.

Indicator	Colour in neutral solution	Colour in base
Litmus	Purple	Blue
Bromothymol blue	Green	Blue
Phenolphthalein	Rose pink	Pink
Methyl orange	Orange-red	Orange
Universal indicator	Green	Blue

Table 4

2.4 Conducting electricity

All alkalis diluted in water form OH⁻ ions. These ions are charge carriers in solutions.
- Aqueous solutions of bases conduct electricity.

2.5 pH scale

- The pH of an alkaline solution is greater than 7.
- As the concentration of OH⁻ ions of an alkali increases, its pH also increases.

2.6 Reaction of bases with acids:

During investigation of the reactions between dilute acids and alkalis in the presence of an indicator, the following are observed:

Fig. 7   The neutralisation of an acid and an alkali and recovering the salt which is one of the products

- The temperature increases until the colour of the indicator changes.
- The point when the solution is completely neutralised, is called the **end point**.
- This reaction is called a neutralisation reaction.
- The products that form, are a salt and water.
- Each acid and base form its own unique salt. When these salts are dissolved in water, solutions with different pH can form. Some salts are neutral like $NaC\ell$, others are acidic like ammonium salts, and others are basic like the carbonates.
- The process of **neutralisation** of acids with alkalis in the laboratory is called **titration**.
- Base + acid → salt + water.

Examples of reactions:

$$NaOH \quad + \ HC\ell \quad\quad \rightarrow \ NaC\ell \ + \ H_2O$$
$$Na_2CO_3 \quad + \ 2HC\ell \quad\quad \rightarrow \ 2NaC\ell \ + \ CO_2 \ + \ H_2O$$
$$Mg(OH)_2 \ + \ (COOH)_2 \ \rightarrow \ Mg(COO)_2 \ + \ 2H_2O$$

When acids react with bases and alkalis

- The acidic properties of acids are destroyed by the alkalis.
- The alkaline properties of bases are destroyed by the acids.
- Salt and water are always products of the reactions.

---

3. Salts

3.1 Preparation of salts

- Salts are ionic compounds consisting of a cation and an anion.
- Salts are formed as products of the following reactions:

General reactions:

Base	+ acid → salt + water
metal	+ acid → salt + hydrogen
metal oxide	+ acid → salt + water
a carbonate	+ acid → salt + water + carbon dioxide
metal hydroxide	+ acid → salt + water

Obtaining the salt crystals from the solutions

- Filter the reaction mixture.
- Evaporate most of the water over a water bath.
- Stop evaporation when tiny crystals begin to form on the sides.
- Keep it in a dust-free place.
- Remove the crystals after a few days.

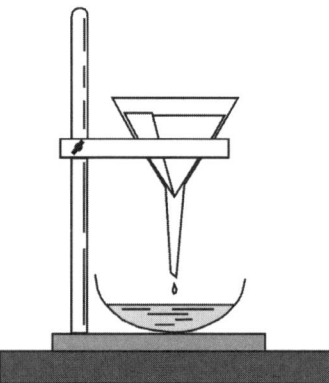

Fig. 8   Filtering the mixture . Recovery of crystals from solution.

3.2 Naming the salts

- The name of the salt is taken from the metal concerned and the anion that is formed by the acid.

  - HCℓ always forms chloride salts
  - HNO$_3$ always forms nitrate salts
  - H$_2$SO$_4$ always forms sulphate salts

- In some cases metals do not react with HNO$_3$ to form nitrate salts and water (when it reacts with metals with strong reducing properties).

QUESTIONS

Section A

I Various possibilities are suggested as answers to the following questions. Indicate the correct answer.

1. Which of the following acids is **not** a mineral acid?
   - A   Hydrochloric acid
   - B   Nitric acid
   - C   Ethanoic acid
   - D   Sulphuric acid

2. When lemon juice is added to an indicator, ...
   - (i) red litmus turns blue.
   - (ii) pink phenolphthalein turns colourless.
   - (iii) universal indicator turns red.
   - (iv) the colour of both methyl-orange and bromothymol blue is red.

   Which statements are true?
   - A   (i) and (ii)
   - B   (ii) and (iv)
   - C   (iii) and (iv)
   - D   (ii) and (iii)

3. Which of the following metals liberate hydrogen gas from an acid?
   - A   Zinc
   - B   Copper
   - C   Silver
   - D   Mercury

4. Which of the following will not form H_3O^+ ions when diluted with water?
   - A H_2CO_3
   - B NH_3
   - C CH_3COOH
   - D $HC\ell$

5. Which indicator is green in a neutral solution?
   - A   Litmus
   - B   Bromothymol blue
   - C   Methyl orange
   - D   Phenolphthalein

6. Which statement is true when the pH of a solution changes from 12 to 2.
   - A   Colourless phenolphthalein turns pink.
   - B   Red litmus turns blue.
   - C   Orange methyl-orange turns red.
   - D   Yellow bromothymol blue turns green.

7. The element present in all acids, is . . .
   - A   oxygen.
   - B   carbon dioxide.
   - C   nitrogen.
   - D   hydrogen.

8. When metals react with acids, ... is always one of the products.
   - A   H
   - B O_2
   - C CO_2
   - D H_2

9. Which of the following general reaction equations is **incorrect**?
   - A   Metal oxide + acid → salt + water
   - B   All metals + acid → salt + water
   - C   Metal hydroxide + acid → salt + water
   - D   Metal carbonate + acid → water + salt + carbon dioxide

10. Which equation correctly shows the dissolution process of ammonia?
    - A $NH_2 + H_2O → NH_3^+(aq) + OH^-(aq)$
    - B $NH_3 + H_2O → NH_4^+(aq) + OH(aq)$
    - C $NH_4^+ \xrightarrow{H_2O} NH_4^+(aq) + OH^-(aq)$
    - D $NH_3 \xrightarrow{H_2O} NH_4^+(aq) + OH^-(aq)$

11. Which of the following will always produce a salt when reacting with an acid?
  (i) Metal   (ii) Alkali   (iii) Metal oxide
  (iv) Non-metal oxide   (v) Carbonate   (vi) Non-metal

  A (ii); (iii) and (v).   C (ii); (iii) and (vi).
  B (i); (ii); (iii) and (v).   D ˙ (iii); (v) and (vi).

12. A compound that contains both an anion and a cation and has a pH of 7 in solution is . . .
  A water.   C an acid.
  B a salt.   D a base.

13. In which of the following possible reactions will a salt be formed?
  A An acid + a carbonate   C A metal + a non-metal
  B A non-metal + an acid   D A base + a metal

14. Which one of the following is a common household acid?
  A $HC\ell$ C H_2SO_4
  B CH_3COOH D $(COOH)_2$

15. The reaction that can be used to prepare a chloride salt is . . .
  A copper + hydrochloric acid.   C copper sulphate + nitric acid.
  B sulphur oxide + hydrochloric acid. D sodium carbonate + hydrochloric acid.

16. Tartaric acid can be regarded as a weak acid because it . . .
  A is usually greatly diluted with water.
  B has a low pH.
  C can react spontaneously with alkalis.
  D is only partially dissociated in water.

17. The pH of a sodium hydroxide solution will probably be . . .
  A 13.   C 7.
  B 8.   D 1.

18. Spilled sulphuric acid is neutralised with sodium carbonate. This will cause the . . .
  A formation of a precipitate.   C formation of a gas.
  B formation of hydrogen.   D pH to go down to 1.

19. Which ion will always be formed when bases dissolve in water?
  A $H_3O^+(aq)$ C $NH_4^+(aq)$
  B $OH^-(aq)$ D $SO_4^{2-}(aq)$

20. Which of the following is **not** an alkali in solution?
  A Ammonia   C NH_4OH
  B NaOH   D CuO

21. An oxide that does not form an alkali in water, is . . .
  A MgO.   C SO_2.
  B Na_2O.   D K_2O.

22. The solution with the highest concentration of H_3O^+ ions is the one with a pH of . . .
  A 14.   C 4.
  B 8.   D 1.

23. The pH of a solution of table salt is . . .
  A smaller than 7.   C equal to 7.
  B much greater than 7.   D slightly greater than 7.

24. Dilute nitric acid and potassium oxide react and form . . .

A   water + carbon dioxide.   C   hydrogen + potassium nitrate.

B   water + potassium nitrate.   D   hydrogen + potassium hydroxide.

25. To prepare sodium carbonate, you need . . .

A   sodium hydroxide + sulphuric acid.

B   sodium carbonate + sulphuric acid.

C   sodium hydroxide + hydrochloric acid.

D   sodium hydroxide + carbonic acid.

26. Which of the following is **not** a general characteristic of an acid?

A   Sour taste   C   Changes the colour of an indicator

B   Soapy feeling   D   pH lower than 7

27. Which statement is **true**?

The end point of a neutralisation reaction is . . .

A   where the pH = 7.

B   where a salt and water are formed.

C   the point where the temperature stabilises.

D   the point where the acid and akali neutralise each other completely.

28. A compound which is insoluble in water and able to destroy an acid, is . . .

A   an alkali.   C $MgCO_3$.

B   NaOH.   D   CaO.

29. Which of the following is true?

A   If you dilute an acid, you change the strength and the pH.

B   If you dilute an acid, you change only the strength.

C   A strong acid produces more H_3O^+ than a weak acid with the same pH.

D   A strong acid produces more H_3O^+ than a weak acid with the same concentration.

30. You want to prepare copper sulphate. Which reaction will **not** be successful if you add the reactants?

A $Cu(OH)_2$ + dilute H_2SO_4 C CuO + dilute H_2SO_4

B Cu + dilute H_2SO_4 D $CuCO_3$ + dilute H_2SO_4

31. The pH of a certain solution is 9. What will happen if a few drops of hydrogen sulphate are added?

A   pH increases; H_3O^+ concentration decreases

B   pH increases; H_3O^+ concentration increases

C   pH decreases; OH^- concentration increases

D   pH decreases; OH^- concentration decreases

II Write down the letter of the term in column B which best suits the description in A. You are not allowed to use a letter more than once.

A	B
1. An example of a weak acid.	A. Neutralisation
2. Indicates the degree of acidity or alkalinity.	B. Strength of acid or base
3. Will not form H_2 in a reaction with an acid.	C. pH = 7
4. Colour of phenolphthalein in a neutral solution.	D. Lime water turns milky
	E. Titration
5. The process by which an acid is added to neutralise a base.	F. pH scale
6. Bases that dissolve in water.	G. This reaction occurs spontaneously.
7. Mg^{2+} and Cu^{2+} are two of them.	H. Standard solution
8. The acidic properties are destroyed when an excess is added to an acid.	I. Burning match causes explosive sound
	J. Mg
9. A base but not a alkali.	K. Form nitrate salts
10. In the presence of H_2O.	L. Purple
11. The test for CO_2.	M. H_3O^+
12. Degree of dissociation.	N. End point
13. Nitric acid and carbonates.	O. Alkalis
14. Will always form when bases are dissolved in water.	P. Cu
	Q. pH < 7
15. Reaction between an alkali and an acid.	R. OH^-
	S. Rose pink
	T. Cations
	U. Vinegar
	V. CuO
	W. Metal oxide
	X. Blue cobalt chloride paper turns pink
	Y. Indicator
	Z. Endothermic

III Read all the statements and complete the table by crossing the best option in each column.

A   Aqueous solutions conduct electricity.
B   Feel soapy.
C   Change the colour of indicators.
D   If you add it to pure water, the pH will increase.
E   Always contain hydrogen.

	A & B	A & C	A & D	A & E	A, B, C & D	A, C & E	B & D	E
1 Acids								
2 Bases								
3 Acid and bases								

Section B

1.   You have to dilute a concentrated acid.

1.1   Describe your method, giving reasons.

1.2   Is the reaction endothermic or exothermic? Why?

2.   Name three characteristics that acids and bases have in common.

3.   Name three characteristics that distinguish between acids and bases.

4.   What is the difference between weak and strong acids and bases? Refer to dissociation and pH.

5.   Which element is responsible for the acidic properties of acids? Explain.

6.   Make critical comments on the following statement:
"Not all bases are alkalis, but all alkalis are definitely bases."

7.   Complete the following general equations:

7.1   metal + acid

7.2   metal oxide + acid

7.3   metal hydroxide + acid

7.4   metal carbonate + acid

7.5   base + acid

8.   Make critical comments on the following statement:
"An acid will always react with a metal to form hydrogen and a salt."

9.   You are provided with five solutions of the same concentration:

Solution	pH
A	14
B	7
C	4
D	9
E	3

9.1 Which solution is the most alkaline?

9.2 Which solution indicates a strong acid?

9.3 Which solution is probably pure water?

9.4 Which one indicates a weak base?

9.5 Which one has the highest concentration H_3O^+ ions?

9.6 Which one will change blue litmus red?

Give reasons for all answers.

10. Criticise the following statement:
"You can change a weak acid like vinegar into a strong acid by adding more acid and increasing the concentration."

11. You have to determine whether a colourless solution is an acid. Describe three different methods you could possibly use.

12. The following elements and compounds are available: Cu; Mg; $HC\ell$; $CaCO_3$; HNO_3; CuO; H_2SO_4.

Give balanced chemical equations to describe how you would prepare each of the following:

12.1 A cylinder filled with hydrogen. How would you prove your success?

12.2 A cylinder filled with carbon dioxide. How would you prove that it is carbon dioxide?

12.3 Copper sulphate crystals. What is the colour of a solution of copper sulphate? Why?

13. Write balanced equations for the following reactions:

13.1 Sodium hydroxide + sulphuric acid

13.2 Calcium carbonate + nitric acid

13.3 Iron + hydrochloric acid

13.4 Copper oxide + hydrogen carbonate

13.5 Potassium hydroxide + ethanoic acid

14. Zinc chloride can be prepared in four different ways, each with an acid and one other reactant. Write balanced equations showing how zinc chloride can be prepared by using

14.1 a hydroxide.

14.2 a metal.

14.3 a carbonate.

14.4 an oxide.

15. Name the acid and metal oxide you would use to prepare sodium nitrate. Give a balanced equation for the chemical reaction.

16. Green copper carbonate powder is added in excess to dilute hydrochloric acid.

16.1 Before the powder is added, universal indicator is added to the solution. What is the colour of the solution?

16.2 Write down your observation after the powder is added, giving reasons.

16.3 Write down a balanced equation for the reaction.

16.4 After the reaction has reached completion, the mixture is filtered. The filtrate is evaporated.

    (a) What is the colour of the filtrate?

    (b) What is the colour of the final powder? Why?

17. Explain how you would prove that marble contains carbonate.

18. You want to prepare table salt by using a neutralisation reaction.

18.1 Which acid would you use? Why?

18.2 Which base would you use?

18.3 Is this base an alkali? Why?

18.4 Which indicator will be used?

18.5 What is the colour of the indicator in

    (a) an acid?

    (b) a base?

    (c) a neutral solution?

18.6 Is the reaction endothermic or exothermic?

18.7 What do we call this method?

18.8 Name the products of the reaction.

18.9 How would you prove the existence of the above neutralisation products?

18.10 Give a balanced equation for the reaction.

19. Complete the following equations:

19.1 $A\ell_2O_3 + HC\ell$

19.2 $Mg(OH)_2 + H_3PO_4$

19.3 $CaCO_3 + HNO_3$

19.4 $Zn + H_2CO_3$

20. Mention five methods to prepare salts.

21. You have prepared solutions of **calcium hydroxide, dilute sulphuric acid, potassium carbonate** and **table salt.** The bottles are not labeled. How would you identify the solutions if you have only bromothymol blue and test tubes available? Write balanced equations to explain.

ANSWERS

Section A

I

1. C	2. D	3. A	4. B	5. B	6. C	7. D
8. D	9. B	10. D	11. A	12. B	13. A	14. B
15. D	16. D	17. A	18. C	19. B	20. D	21. C
22. D	23. C	24. B	25. D	26. B	27. D	28. C
29. D	30. B	31. D				

II

1. U	2. F	3. P	4. S	5. E	6. O	7. T
8. W	9. V	10. X	11. D	12. B	13. K	14. R
15. A						

III

	A & B	A & C	A & D	A & E	A, B, C & D	A, C & E	B & D	E
1 Acids				X		X		X
2 Bases	X		X		X		X	
3 Acids and bases		X						

Section B

1.1 Diluting acids
- The concentrated acid must be added to water
- Add a little acid at a time, and stir all the time – the reaction produces heat.
- Stick labels to containers of diluted acids – to avoid confusion

1.2 Exothermic. Heat is released. Temperature rises.

2. Both:
- Change the colour of indicators
- Conduct electricity
- Change the pH if added to pure water
- Destroy each other.

3.	Acids	Bases
• When diluted in water . . . is formed:	H_3O^+	OH^-
• When added to pure water: pH Concentration H_3O^+	decreases increases	increases decreases
• Change the colour: Litmus Bromothymol blue Phenolphthalein Methyl orange Universal indicator	red yellow colourless red red	blue blue pink orange blue
• Taste and feel	taste sour	feel soapy

4.   Strong acids and bases:

- Strong acids and bases dissociate completely
- The pH depends on the number of H_3O^+ ions
- If you add 1 teaspoon of $HC\ell$ to a litre of water, it will dissociate completely. All the $HC\ell$ will form H_3O^+ ions and $C\ell^-$, no $HC\ell$ left – strong acid
- The pH will be between 1 and 3 because the concentration of H_3O^+ ions is high.
- $HC\ell$ is a strong acid

Weak acids and bases:

- Weak acids and bases dissociate incompletely
- The pH will be between 3 and 9 because the concentration of H_3O^+ ions is low.
- If you add 3 cups of vinegar to 1 litre of water, it will ionize partially. Not all the CH_3COOH will form $CH_3COO^- + H_3O^+$; CH_3COOH will be left in the solution – weak acid

5.   Hydrogen. When an acid is added to water, it forms H_3O^+ ions
- $H^+ + H_2O \rightarrow H_3O^+$

6.   It is true
- Not all bases are soluble in water
- Bases that are soluble are called alkalis
- All bases are therefore not alkalis
- All alkalis are definitely bases

7.1	→ hydrogen + salt (if metal is more reactive than hydrogen) or no reaction (if metal is less reactive than hydrogen)
7.2	→ salt + water
7.3	→ salt + water
7.4	→ salt + water + carbon dioxide
7.5	→ salt + water

8. False
- It only happens if the metal is more reactive than hydrogen (see section on "Reactions of metals") If nitric acid is used, the acid must be in a very diluted state.

9.1 A
- highest pH
- lowest H_3O^+ concentration
- highest OH^- concentration

9.2 E
- pH $<$ 7
- because they all have the same concentration, but E has the most H_3O^+ ions
- this indicates that E dissociates more completely than C.

9.3 B
- pH $=$ 7
- same amount H_3O^+ ions and OH^- ions

9.4 D
- pH $>$ 7
- because they all have the same concentration, but D has less OH^- than A
- this indicates that D dissociates less than A

9.5 E
- Lowest pH; most H_3O^+ ions

9.6 E, C
- pH $<$ 7
- acids
- acids change the colour of blue litmus to red.

10. False
- weak acids are weak because they dissociate partially
- by adding more vinegar you will only change the concentration
- pH will decrease because concentration of H_3O^+ ions increases
- strength remains the same

11.1 Add an indicator
- If the colour changes according to Table 2, it is an acid.

11.2 Add a carbonate
- If a gas is liberated
- Test the gas
- If it turns clear lime water milky
- It is CO_2
- The solution is an acid

11.3 Add a metal more reactive than hydrogen
- If a gas is liberated
- Test the gas
- If it ignites with an explosive sound
- It is hydrogen
- The solution is an acid

12.1 $Mg + 2HC\ell \rightarrow MgC\ell_2 + H_2$
  * Test the gas with a burning match
  * an explosive sound will be heard

12.2 $CaCO_3 + 2HNO_3 \rightarrow Ca(NO_3)_2 + H_2O + CO_2$ ($CaCO_3 + HC\ell$ can also be used)
  * Test the gas with clear lime water
  * The clear lime water turns milky

12.3 $CuO + H_2SO_4 \rightarrow CuSO_4 + H_2O$
  * Blue. Because of the Cu^{2+} ions and the presence of water
    (hydrated Cu^{2+} ions)

13.1 $2NaOH + H_2SO_4 \rightarrow Na_2SO_4 + 2H_2O$

13.2 $CaCO_3 + 2HNO_3 \rightarrow Ca(NO_3)_2 + H_2O$

13.3 $Fe + HC\ell \rightarrow FeC\ell_2 + H_2$

13.4 $CuO + H_2CO_3 \rightarrow CuCO_3 + H_2O$

13.5 $KOH + CH_3COOH \rightarrow KCH_3COO + H_2O$

14.1 $Zn(OH)_2 + 2HC\ell \rightarrow ZnC\ell_2 + 2H_2O$

14.2 $Zn + 2HC\ell \rightarrow ZnC\ell_2 + H_2$

14.3 $ZnCO_3 + 2HC\ell \rightarrow ZnC\ell_2 + CO_2 + H_2O$

14.4 $ZnO + 2HC\ell \rightarrow ZnC\ell_2 + H_2O$

---

15. HNO_3 and Na_2O
  $Na_2O + 2HNO_3 \rightarrow 2NaNO_3 + H_2O$

16.1 Red

16.2 After the powder is added:
  • A gas is liberated (carbonate forms CO_2)
  • The amount of green powder decreases but in the end there is still green
    powder left ($CuCO_3$ was in excess)
  • The colour of the indicator changes from red to green and later to blue-green.
    ($CuCO_3$ destroys the acidic properties, becomes neutral; later the $CuCO_3$ in excess
    changes the colour to blue, but copper carbonate is less soluble in water.)

16.3 $CuCO_3 + 2HC\ell \rightarrow CuC\ell_2 + CO_2(g) + H_2O$

16.4 (a) Blue
  • The presence of Cu^{2+} and water

  (b) Green
  • $CuC\ell_2$ is green
  • No water of crystallisation

---

17. • Add any acid to the granules
  • If a gas is liberated. Test the gas.
  • If it turns clear lime water milky, it is carbon dioxide (CO_2)
  • The granules definitely contain carbonate ions
  • A carbonate + acid $\rightarrow$ salt + H_2O + $CO_2(g)$

18.1 HCℓ
  - HCℓ gives chloride salts

18.2 NaOH or Na$_2$O
  - Table salt is a compound containing sodium and chloride ions.

18.3 Alkali
  - A base that is soluble in water

18.4 Bromothymol blue

18.5 (a) Yellow

  (b) Blue

  (c) Green

18.6 Exothermic
  - Heat is released during the reaction

18.7 Titration

18.8 salt + water
  NaCℓ + H$_2$O

18.9 H$_2$O
  - Test with blue cobalt chloride paper, it turns pink
  - Or white water-free (anhydrous) CuSO$_4$, powder becomes blue
  NaCℓ
  - Evaporates it
  - Taste the salt

18.10 NaOH + HCℓ → NaCℓ + H$_2$O
  Na$_2$O + 2HCℓ → 2NaCℓ + H$_2$O

19.1 Aℓ_2O$_3$ + 6HCℓ → 2AℓCℓ_3 + 3H$_2$O

19.2 3Mg(OH)$_2$ + 2H$_3$PO$_4$ → Mg$_3$(PO$_4$)$_2$ + 6H$_2$O

19.3 CaCO$_3$ + 2HNO$_3$ → Ca(NO$_3$)$_2$ + CO$_2$(g) + H$_2$O

19.4 Zn + H$_2$CO$_3$ → ZnCO$_3$ + H$_2$(g)

20.
  - metal + acid → salt + H$_2$
  - metal oxide + acid → salt + H$_2$O
  - metal hydroxide + acid → salt + H$_2$O
  - carbonate + acid → salt + H$_2$O + CO$_2$
  - base + acid → salt + H$_2$O

---

21. Test all the solutions with bromothymol blue
  - Only the sulphuric acid will change the colour to yellow
  - Label the bottle
  - The table salt will change the colour to green
  - Label the bottle
  Add the sulphuric acid to both the other solutions
  - The one that liberates a gas, CO$_2$, is potassium carbonate
  - Label the bottle potassium carbonate
  - A carbonate + an acid → salt + H$_2$O + CO$_2$
  - Label the remaining bottle calcium hydroxide

11 Ions, Electrolytes and Electrochemical Reactions

This chapter is concerned with the ways in which an electric current moves through different substances. In previous chapters you have studied the nature of an electric current. Electrons act as current carriers in metallic conductors. In this chapter we are going to look at conductors of current in solutions and at the possibility that the addition of electrical energy can cause a chemical reaction.

1. Evidence of the electrical conductivity of chemical substances

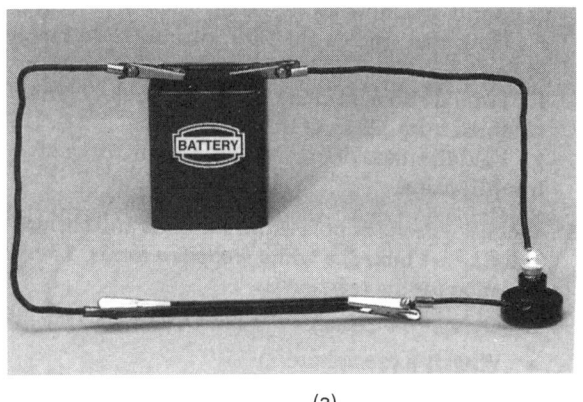
(a)

(b)

Fig. 1   Apparatus for investigating the conductivity of (a) solids and (b) solutions.

Results of such an investigation:
- All solid substances are insulators except for metals and carbon.
- Solutions of ionic solids conduct an electric current.
- Covalent liquids do not conduct an electric current.
- The solid ionic crystals do not conduct an electric current because the ions are firmly held in position by electrostatic forces.
- Most salts dissolve in water to form ions.

Example reaction:

$$NaC\ell \overset{\text{water}}{\underset{\text{dissolves}}{\to}} Na^{1+} (aq) + C\ell^{1-} (aq)$$

$\left[\text{Ionic changes of } 1^{+} \text{ and } 1^{-} \text{ are from now on indicated by } + \text{ and } - \text{ respectively.}\right]$

- When the crystals dissolve in water, the forces are disrupted and the ions become free to act as current carriers.
- Purified water does not conduct electricity.
- Solutions of acids, bases and salts also conduct electricity.
- Most salts dissolve in water to form ions. If they conduct an electric current, they are called **electrolytes**.
- Substances that do not form ions and do not conduct an electric current, are called **non-electrolytes.**

2. Ions and ionic charges

2.1 Ions consisting of one atom only

- The metals in group 1 of the Periodic Table releases **one electron** to form ions with a positive charge of +1.
  - $* \ Li \rightarrow Li^+ + e^-$
  - $* \ Na \rightarrow Na^+ + e^-$
  - $* \ K \rightarrow K^+ + e^-$

- The metals in group II of the Periodic Table release **two electrons** to form ions with a positive charge of +2.
  - $* \ Mg \rightarrow Mg^{2+} + 2e^-$
  - $* \ Ca \rightarrow Ca^{2+} + 2e^-$

- The metals in group III of the Periodic Table release **three electrons** to form ions with a positive charge of +3.
  - $* \ A\ell \rightarrow A\ell^{3+} + 3e^-$

- The non-metals in group VII of the Periodic Table gain **one electron** to form negative ions with a charge of –1.
  - $* \ F_2 + 2e^- \rightarrow 2F^-$
  - $* \ C\ell_2 + 2e^- \rightarrow 2C\ell^-$
  - $* \ Br_2 + 2e^- \rightarrow 2Br^-$

- The non-metals in group VI of the Periodic Table gain **two electrons** to form negative ions with a charge of –2.
  - $* \ O + 2e^- \rightarrow O^{2-}$
  - $* \ S + 2e^- \rightarrow S^{2-}$

- The non-metals in group V of the Periodic Table gain **three electrons** to form negative ions with a charge of –3.
  - $* \ N + 3e^- \rightarrow N^{3-}$

2.2 Ions consisting of groups of atoms

- Groups of atoms that are bonded together covalently, may also have a deficiency or an excess of electrons. **The most common ones are included in Table 8.2**.

3. Electrolysis

3.1 The electrolysis of a molten ionic compound

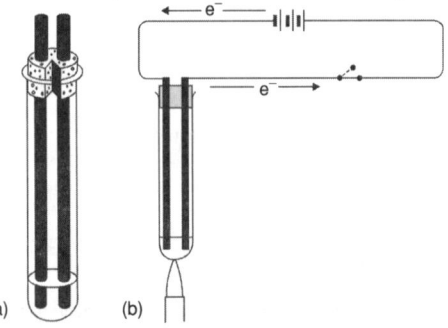

(a) (b)

Fig. 2   The electrolysis of molten lead bromide; (a) the electrolytic cell and (b) the circuit diagram

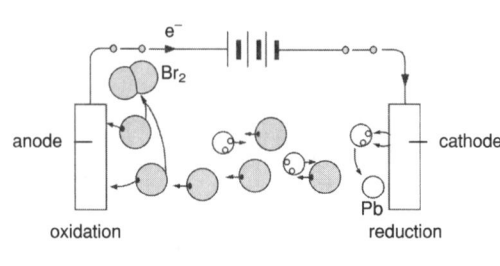

anode   oxidation   Pb   cathode   reduction   Br$_2$

Fig. 3   Schematic representation of the electrolysis of lead bromide

- Solid lead(II) bromide does not conduct an electric current.
- Molten lead(II) bromide does conduct an electric current.
- The energy transferred to the lead bromide overcomes the electrostatic forces.
- The phase change allows the ions to act as current carriers.

$$PbBr_2 \underset{\Delta}{\rightarrow} Pb^{2+} + 2\ Br^-$$

- Lead deposits on one rod:

$$Pb^{2+} + 2e^- \rightarrow Pb$$

- Brown bromine fumes are produced at the other rod:

$$2\ Br^- \rightarrow Br_2 + 2e^-$$

- The net reaction is represented by

$$PbBr_2 \rightarrow Pb + Br_2$$

- The electrical energy transferred to the lead bromide is converted into chemical energy.

- The current flows in two parts of the electrolytic cell; through the molten lead bromide and the external circuit. In the molten lead the current is carried by ions and in the external circuit the free electrons in the metal conductors carry the current.

- The **negative electrode** (rod) is called the **cathode**.

  * The cathode attracts the positive ions.
  * There is an excess of electrons on the cathode.
  * The electrode transfers the electrons to the positive ions and a metal deposits.
  * **Electrons** are **gained** here and **reduction** occurs.

- The **positive electrode** (rod) is called the **anode**.

  * The anode attracts the negative ions.
  * There is a deficiency of electrons on the anode.
  * The negative ions transfer electrons to the anode and a non-metal is formed.
  * **Electrons** are **liberated** here and **oxidation** occurs.

- The electrons released at the positive electrode move around to the negative electrode. They are then transferred to the positive ions.

3.2 Electrolyses of a concentrated solution of copper(II) chloride

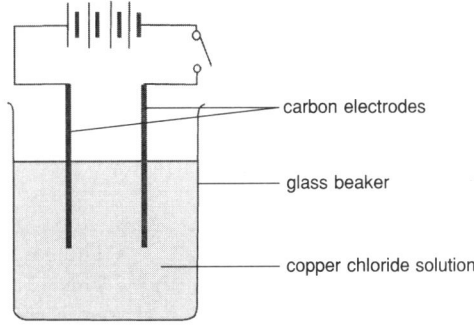

Fig. 4   The electrolysis of a copper(II) chloride solution

193

- The copper chloride dissolves in the water

$$CuC\ell_2 \rightarrow Cu^{2+} + 2\ C\ell^-$$

- On the **cathode** which is connected to the negative terminal of the battery, a reddish brown deposit forms.

$$Cu^{2+} + 2e^- \rightarrow Cu$$

- A yellowish-green gas collects at the **anode** which is connected to the positive terminal of the battery.

$$2C\ell^- \rightarrow C\ell_2 + 2e^-$$

- The net reaction is represented by
  $$CuC\ell_2 \rightarrow Cu + C\ell_2$$

- The current flows in two parts of the electrolytic cell; through the copper chloride solution and the external circuit. The current carriers are ions and electrons in the solution and the metal conductor respectively.

3.3 Practical applications of electrolysis

- Electroplating

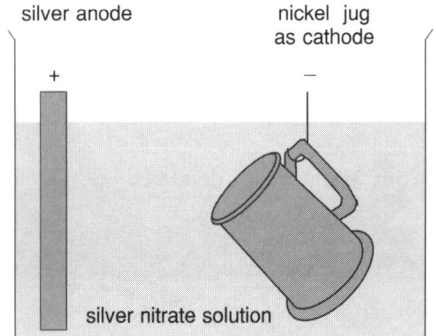

silver anode   nickel jug as cathode

+   −

silver nitrate solution

(a) The electrolytic cell with the nickel jug as cathode   (b) The silverplated jug.

Fig. 5   Electroplating

- Preparation of chemicals

- Extraction of metals.

4. Reactions producing electrical energy

4.1 Redox reactions

- During a redox reaction electron transfer occurs.
- A redox reaction consists of two half reactions, namely:
- **Oxidation** and **reduction** which take place simultaneously.

Oxidation	Reduction
• The loss of electrons by a substance	• The gain of electrons by a substance
• The reducing agent undergoes oxidation and loses electrons.	• The oxidising agent undergoes reduction and gains electrons.
• Occurs at the anode of the cell.	• Occurs at the cathode of the cell.

The **only valid definitions** for **cathode** and **anode** are those **in terms of oxidation and reduction**. The **cathode** is that elctrode where **reduction** occurs.
The **anode** is that electrode where **oxidation** occurs.
In the electrolitic cell, the cathode is the negative and the anode the positive electrode. In the voltaic cell, however, it is the other way round. To avoid confusion, we never use the rule that the anode is the positive and the cathode the negative electrode.

4.2 Relative strength of oxidising and reducing agents (Refer to Table 11.1)

- **In a redox reaction** competition for electrons exists.
- Some substances acquire electrons easier than others.
- Ag^+ takes up electrons more readily than Cu^{2+}, which takes up electrons more readily than Mg^{2+}
- The **substance that takes up electrons the easiest**, is the **strongest oxidising agent.**
- The substances on the left of the double arrows are the **oxidising agents.**
- The substance on the right of the double arrows are the **reducing agents.**
- **Reduction half reactions** are **read from left to right.**
- **Oxidation half reactions** are **read from right to left.**
- An oxidation-reduction reaction will proceed spontaneously if a substance on the right reacts with a substance on the left which appears below it on the Table.
- If the circle is **anti-clockwise,** the **reaction is spontaneous.**

Example

Copper metal is added to a solution of $AgNO_3$. Cu occurs on the right hand side of the table above the oxidising agent Ag^+. Therefore the reaction will proceed spontaneously.

$$Cu + 2\,Ag^+ \rightarrow Cu^{2+} + 2\,Ag$$

4.3 Direct contact

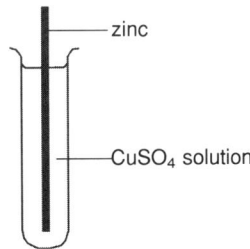

Fig. 6   Direct contact between zinc and a copper sulphate solution.

- The temperature rises.
- The reaction is exothermic. Energy is liberated.
- The copper sulphate solution becomes colourless.
- No copper ions are left in the solution.
- Brown copper is deposited on the zinc.
- Zinc ions are formed in the solution.

Table with relative strengths of oxidising and reducing agents

	Half-reactions	
	Oxidising agents Reducing agents	
Weakest oxidising agents	$Li^+ + e^- \rightleftharpoons Li$ $K^+ + e^- \rightleftharpoons K$ $Cs^+ + e^- \rightleftharpoons Cs$ $Ba^{2+} + 2e^- \rightleftharpoons Ba$ $Sr^{2+} + 2e^- \rightleftharpoons Sr$ $Ca^{2+} + 2e^- \rightleftharpoons Ca$ $Na^+ + e^- \rightleftharpoons Na$ $Mg^{2+} + 2e^- \rightleftharpoons Mg$ $A\ell^{3+} + 3e^- \rightleftharpoons A\ell$ $Mn^{2+} + 2e^- \rightleftharpoons Mn$ $2H_2O + 2e^- \rightleftharpoons H_2 + 2OH^-$ $Zn^{2+} + 2e^- \rightleftharpoons Zn$ $Cr^{3+} + 3e^- \rightleftharpoons Cr$ $Fe^{2+} + 2e^- \rightleftharpoons Fe$ $Co^{2+} + 2e^- \rightleftharpoons Co$ $Ni^{2+} + 2e^- \rightleftharpoons Ni$ $Sn^{2+} + 2e^- \rightleftharpoons Sn$ $Pb^{2+} + 2e^- \rightleftharpoons Pb$ $Fe^{3+} + 3e^- \rightleftharpoons Fe$ $\mathbf{2H^+ + 2e^- \rightleftharpoons H_2}$ $S + 2H^+ + 2e^- \rightleftharpoons H_2S$ $Sn^{4+} + 2e^- \rightleftharpoons Sn^{2+}$ $Cu^{2+} + e^- \rightleftharpoons Cu^+$ $SO_4^{2-} + 4H^+ + 2e^- \rightleftharpoons SO_2 + 2H_2O$ $Cu^{2+} + 2e^- \rightleftharpoons Cu$ $2H_2O + O_2 + 4e^- \rightleftharpoons 4OH^-$ $Cu^+ + e^- \rightleftharpoons Cu$ $I_2 + 2e^- \rightleftharpoons 2I^-$ $O_2 + 2H^+ + 2e^- \rightleftharpoons H_2O_2$ $Fe^{3+} + e^- \rightleftharpoons Fe^{2+}$ $Hg^{2+} + 2e^- \rightleftharpoons Hg$ $NO_3^- + 2H^+ + e^- \rightleftharpoons NO_2 + H_2O$ $Ag^+ + e^- \rightleftharpoons Ag$ $NO_3^- + 4H^+ + 3e^- \rightleftharpoons NO + 2H_2O$ $Br_2 + 2e^- \rightleftharpoons 2Br^-$ $MnO_2 + 4H^+ + 2e^- \rightleftharpoons Mn^{2+} + 2H_2O$ $O_2 + 4H^+ + 4e^- \rightleftharpoons 2H_2O$ $Cr_2O_7^{2-} + 14H^+ + 6e^- \rightleftharpoons 2Cr^{3+} + 7H_2O$ $C\ell_2 + 2e^- \rightleftharpoons 2C\ell^-$ $Au^{3+} + 3e^- \rightleftharpoons Au$ $MnO_4^- + 8H^+ + 5e^- \rightleftharpoons Mn^{2+} + 4H_2O$ $H_2O_2 + 2H^+ + 2e^- \rightleftharpoons 2H_2O$ $F_2 + 2e^- \rightleftharpoons 2F^-$	Strongest reducing agents ↑
Strongest oxidising agents		Weakest reducing agents

————————————— REDUCTION ————————————→

←———————————— OXIDATION —————————————

Table 11.1   Oxidation and reduction half reactions

The use of the Table is not prescribed for Grade 10 learners. The only redox reactions of importance in this chapter are those involved in the prescribed electrolitic cells and the voltaic cell. You will again come across this Table in Grade 12.

The net reaction for the process above can be obtained from the Table:

Oxidation: (Read from right to left) $Zn \rightarrow Zn^{2+} + 2e^-$

Reduction: (Read from left to right) $Cu^{2+} + 2e^- \rightarrow Cu$

Net redox reaction: (add up) $Zn + Cu^{2+} \rightarrow Cu + Zn^{2+}$

4.4 Indirect contact

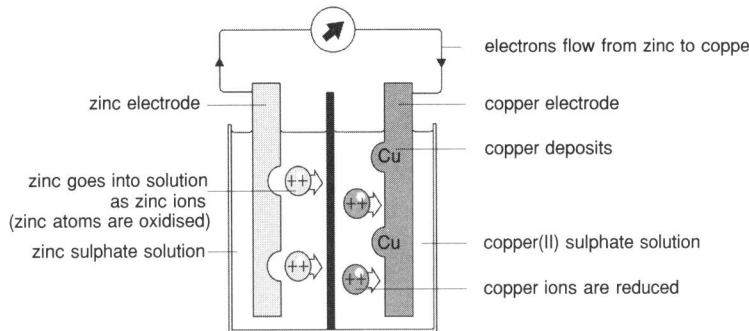

Fig. 7   A complete electrochemical cell – in this case it is a voltaic cell

- Electrons flow through the meter from the Zn electrode to the Cu electrode.
- The Zn loses electrons:

$Zn \rightarrow Zn^{2+} + 2e^-$

- The mass of the Zn electrode decreases.
- A thin, brown layer of copper is deposited on the copper electrode.
- The liberated electrons flow through the external circuit to the copper electrode.
- The copper ions take up the electrons and copper forms:

$Cu^{2+} + 2e^- \rightarrow Cu$

- The mass of the copper electrode increases.
- The reading on the voltmeter is 1,1 V.
- A redox reaction occurs:

$Zn + Cu^{2+} \rightarrow Cu + Zn^{2+}$

- Chemical energy is converted into electrical energy.
- This kind of cell is called a voltaic cell. It produces electrical potential energy.
- Cell notation:

$Zn \mid Zn^{2+} \parallel Cu^{2+} \mid Cu$

To summarise:
There has to be distinguished between two types of reactions:

1. **Reactions that need electrical energy. Electrical energy** is converted into **chemical energy** in the following cells (**electrolytic cells**)
   (a) Molten lead bromide between carbon electrodes.
   (b) A concentrated solution of copper chloride between carbon or platinum electrodes.

2. **Reactions that produce electrical energy spontaneously. Chemical energy** is converted into **electrical energy** in the following cell (**voltaic cell**):
   Zn and Cu electrodes in solutions of Zn^{2+} and Cu^{2+} respectively.

5. Soluble and insoluble salts

- Soluble as well as insoluble salts consist of ions.
- When aqueous solutions of salts are mixed, a precipitate is sometimes obtained and sometimes no reaction occurs.
- A precipitate is obtained when a combination of ions together forms an insoluble salt.

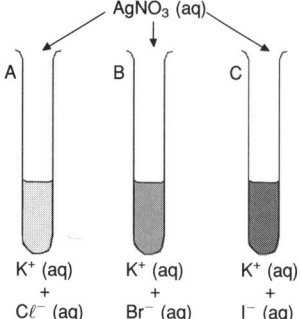

In A: A white AgCℓ precipitate forms

In B: A cream AgBr precipitate forms

In C: A yellow AgI precipitate forms

Fig. 8   Reactions between aqueous salt solutions.

Example 1

Solutions of KNO_3 and $AgNO_3$ are added together:

KNO_3 dissolves in water and $K^+(aq)$ and $NO_3^-(aq)$ are formed.

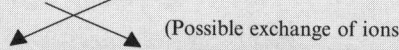

(Possible exchange of ions)

$AgNO_3$ dissolves in water and $Ag^+(aq)$ and $NO_3^-(aq)$ are formed.

Possible new products are KNO_3 + $AgNO_3$. Both are soluble, so no precipitate is formed.

Example 2

Solutions of $Pb(NO_3)_2$ and KI are added together.

$Pb(NO_3)_2$ dissolves in water and $Pb^{2+}(aq)$ and $NO_3^-(aq)$ are formed

(Possible exchange of ions)

KI dissolves in water and $K^+(aq)$ and $I^-(aq)$ are formed.

- The possible new products are KNO_3, which is soluble, and PbI_2, which is insoluble, so a yellow precipitate is formed.
- K^+ and NO_3^- ions are present as spectator ions.

The reaction:

$$Pb(NO_3)_2 + 2\ KI \rightarrow 2\ KNO_3 + PbI_2\ (s)\downarrow$$
$$\text{yellow}$$

5.1 Solubility of salts

A knowledge of the solubility rules are useful for the following reasons:
- To prepare insoluble salts.
- To test for the presence of certain anions.

5.2 Solubility rules

- **All nitrates and acetates are soluble.**
- **All salts of the alkali metals and ammonia are soluble.**
- **All chloride, bromide and iodide salts are soluble except those of silver and lead:**

$AgC\ell$	white
$AgBr$	cream
AgI	light yellow
$PbC\ell_2$	white
$PbBr_2$	light yellow
PbI_2	yellow

To test for the presence of halide ions in solutions, a solution of silver nitrate is added and the specific compounds are then identified by their colour.

- **All sulphates are soluble except those of barium, calcium, lead and silver, eg.**

$BaSO_4$	white
$CaSO_4$	white

To test for a sulphate, a solution of barium chloride is added. A white precipitate indicates that the test is positive.

- **Most carbonates are almost insoluble except those of the alkali metals and ammonia.**

Test for carbonates: Carbonates react with acids to liberate a gas, carbon dioxide. (The gas causes clear lime water to turn milky white)

6. Calculations related to ionic reactions (HG learners only)

- The mass of any reactant or product in a reaction can be calculated by using the relative formula masses of compounds.

Example

Calculate the mass of lead iodide (PbI_2) formed when 31g of $Pb(NO_3)_2$ react in a chemical reaction with potassium iodide.

$$Pb(NO_3)_2 + 2\,KI \rightarrow 2\,KNO_3 + PbI_2\,(s)\downarrow$$
$$\text{yellow}$$

$M_r[Pb(NO_3)_2] = 207 + [14 + 16(3)]\,2 = 331$

$M_r[KI] \qquad = 39 + 127 = 166$

$M_r[KNO_3] \quad = 39 + 14 + (16)3 = 101$

$M_r[PbI_2] \quad = 207 + (127)2 = 461$

(Solution continues on next page)

If you use 331 g $Pb(NO_3)_2$, 461 g yellow PbI_2 (s) will form.

If you use 31 g $Pb(NO_3)_2$ how many grams of yellow PbI_2 (s) will form?

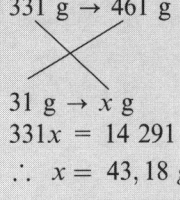

331 g → 461 g

31 g → x g

$331x = 14\ 291$

$\therefore \ x = 43,18\ g$

For additional exercise, see question 13 in Section B of the questions following immediately after this background study.

QUESTIONS

SECTION A

Various possibilities are suggested as answers to the following questions. Choose the correct answer.

1. Which of the following substances will conduct an electric current?
   A $NaC\ell$ crystals   C Purified water
   B Ethanol   D A copper chloride solution

2. The charge carriers in a metallic conductor and an electrolyte are . . .
   A ions and electrons respectively.   C both electrons.
   B electrons and ions respectively.   D both ions.

3. The correct chemical formulae for ammonium sulphate and aluminium oxide are respectively . . .
   A $(NH_4)_2SO_4$ and $A\ell O$.   C NH_4SO_4 and $A\ell_2O_3$.
   B $(NH_4)_2SO_4$ and $A\ell O_2$.   D $(NH_4)_2SO_4$ and $A\ell_2O_3$.

4. Study the diagram of the electrolysis of a concentrated solution of $CuC\ell_2$ (Fig. 4)

4.1 Which substance will form at the negative electrode?
   A Cu   C Cu^{2+}
   B $C\ell_2$ D $C\ell^-$

4.2 Which substance will form at the anode?
   A Cu   C Cu^{2+}
   B $C\ell_2$ D $C\ell^-$

4.3 Which of the following acts as the oxidising agent?
   A Cu   C Cu^{2+}
   B $C\ell_2$ D $C\ell^-$

4.4 The reduction half reaction is represented by . . .
   A $CuC\ell_2 \rightarrow Cu^{2+} + 2C\ell^-$ C $2C\ell^- \rightarrow C\ell_2 + 2e^-$
   B $Cu^{2+} + 2e^- \rightarrow Cu$ D $C\ell_{2+} + 2e^- \rightarrow 2C\ell$

5. An oxidising agent is a substance which . . .
   A accepts electrons.   C accepts protons.
   B donates electrons.   D donates protons.

6. $Zn + 2HC\ell \rightarrow ZnC\ell_2 + H_2$
   What type of reaction does the above equation represent?
   A An ion exchange reaction   C An acid-base reaction
   B An oxidation reaction   D A redox reaction

7. A zinc rod is placed in a blue copper sulphate solution. After a few hours the colour of the solution fades. The reason for this is that the. . .
   A zinc rod oxidises the copper ions.
   B sulphate ions are reduced to sulphur dioxide.
   C copper ions accept electrons to form copper.
   D zinc rod forms zinc ions.

8. In the reaction $Cu(s) + 2 Ag^+ (aq) \rightarrow Cu^{2+} (aq) + 2 Ag(s)$ . . .
   A copper is the oxidising agent.
   B the silver ions are the oxidising agents.
   C copper donates electrons to the silver cation.
   D copper is reduced.

9. In the half-reaction
   $Y^{2-} \rightarrow Y + 2e^-$ the Y^{2-} ion is ... and is called the ...
   A reduced, oxidising agent.
   B oxidised, oxidising agent.
   C oxidised, reducing agent.
   D reduced, reducing agent.

10. Which of the following statements is/are true about a **voltaic cell?**
    (i) The oxidation half-reaction occurs at the cathode of the cell.
    (ii) Positive ions move from the salt bridge to the anode of the cell.
    A Neither (i) nor (ii)   C (ii) only
    B (i) only   D Both (i) and (ii)

11. Which of the following is true about an **electrolytic cell?**
    (i) The oxidation half-reaction occurs at the anode of the cell.
    (ii) The anode is the negative electrode.

    A Neither (i) nor (ii)   C (ii) only
    B (i) only   D Both (i) and (ii)

12. When an atom forms a positive ion , it ...
    A loses electrons and is reduced.
    B gains electrons and is oxidised.
    C gains electrons and is reduced.
    D loses electrons and is oxidised.

13. The information in the Table shows that Cu^{2+} is a stronger oxidising agent than ...
    A Zn^{2+}.   C Ag^+.
    B Hg^{2+}.   D Au^{3+}.

14. The correct cell notation for the zinc-copper electrochemical cell is ...
    A $Zn^{2+} \mid Zn \parallel Cu^{+2} \mid Cu$.
    B $Zn \mid Zn^{2+} \parallel Cu \mid Cu^{2+}$.
    C $Zn \mid Zn^{+2} \parallel Cu^{+2} \mid Cu$.
    D $Cu^{+2} \mid Cu \parallel Zn \mid Zn^{2+}$.

15. The stronger the oxidising agent, the ...
    A weaker its ability to gain electrons.
    B stronger its ability to donate electrons.
    C weaker its ability to take part in a reaction.
    D stronger its ability to gain electrons.

16. Mg ribbon is dropped into different solutions. Where will a spontaneous reaction take place?

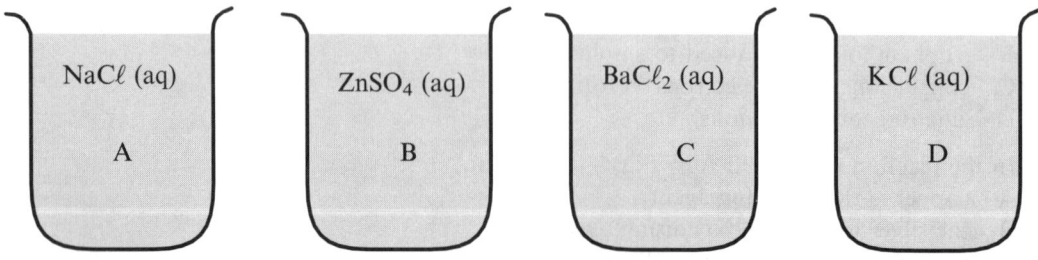

202

17. The anode in an electrochemical cell is . . .
    A always placed on the left-hand side of the cell.
    B the negative electrode.
    C the electrode where oxidation takes place.
    D the electrode where reduction takes place.

18. Which one of the following solutions cannot be stored in a zinc bucket?
    A $MgSO_4$ C $CuSO_4$
    B $A\ell C\ell_3$ D $NaC\ell$

19. A spontaneous reaction occurs when chlorine gas comes into contact with hydrogen gas. The reason for this is probably that . . .
    A chlorine is a good oxidising agent.
    B hydrogen is a good oxidising agent.
    C chlorine is a good reducing agent.
    D hydrogen and chlorine are non-metals.

20. Which of the following substances will reduce lead ions but not zinc ions?
    A $A\ell$ C Fe
    B Ag   D Cu

21. Which of the following is according to the Table the strongest reducing agent?
    A $A\ell$ C Ni
    B Cu^{2+} D Sn^{2+}

22. A test for a sulphate is conducted and the result is positive. Which substance **cannot** be used during the test?
    A Barium chloride   C Silver nitrate
    B Copper sulphate   D Barium nitrate

23. A precipitate is formed and a gas liberated when hydrochloric acid is added to an unkown substance. The substance is probably . . .
    A silver nitrate.   C lead nitrate.
    B silver carbonate.   D barium chloride.

24. Sodium carbonate and silver nitrate solutions are added together. The spectator ions are . . .
    A Na^+ and NO_3^- C Ag^+ and CO_3^{2-}
    B Na^+ and CO_3^{2-} D Ag^+ and NO_3^-

25. Which one of the following is insoluble in water?
    A Sodium sulphate   C Ammonium carbonate
    B Potassium bromide   D Calcium sulphate

26. Which one of the following salts will effervesce with a dilute acid?
    A Copper oxide   C Ammonium carbonate
    B Sodium chloride   D Lead sulphate

27. Which of the following will **not** cause a precipitate if it is added to a solution of lead nitrate?
    A Silver nitrate   C Copper chloride
    B Sodium chloride   D Potassium chloride

SECTION B

1. Study the following list of substances:

Dilute hydrochloric acid	Paraffin
Purified water	Copper sulphate crystals
Sugar	Sodium chloride solution
Solution of naphthalene in alcohol	Tap water

   Use the list and
   1.1 name three examples of conductors of electric current.
   1.2 name three examples of substances that will not conduct an electric current.
   1.3 identify all the electrolytes.

2. Explain what the term "electrolyte" means.

3. Dissolve sugar and copper sulphate crystals in water. Which solution conducts an electric current? Explain your answer?

4. Why does tap water conduct an electric current but purified water not?

5. What are the particles called that act as current carriers in electrolytes?

6. Criticise the following statement: "All electrolytes are soluble in water." Illustrate your argument with suitable examples.

7. Write down the correct chemical formulae for the following substances:

   7.1 Sodium carbonate
   7.2 Magnesium hydroxide
   7.3 Ammonium chloride
   7.4 Copper nitrate

8. Explain the difference between oxidation and reduction by referring to electron transfer.

9. Explain the difference between electrolytic and voltaic cells. Refer to energy transfer.

10. Distinguish between endothermic and exothermic reactions.

11. Name two applications of electrolysis in practice.

12. A concentrated copper chloride solution undergoes electrolysis.

    12.1 Write down a balanced equation to indicate how the substance dissociates.
    12.2 At which electrode will copper be formed?
    12.3 Write down the reduction half-reaction.
    12.4 Identify the
        (a) oxidising agent.
        (b) the substance that will be oxidised.

13. An excess potassium bromide is added to a solution containing 4,25 g silver nitrate.
    13.1 Calculate the molar formula mass of silver nitrate.
    13.2 Calculate the mass of precipitate that is formed.

14. A copper sulphate solution is added to a solution of calcium chloride. Predict whether any precipitate will form. Prove your answer by balanced equations.

15. Which solutions should be added to solutions of unknown salts to test for the presence of each of the following anions? Write formulae for the precipitates formed.
    1. Chloride
    2. Bromide
    3. Iodide
    4. Sulphate
    5. Carbonate

16. How would you distinguish between two unlabelled bottles, one filled with hydrochloric acid and the other one filled with sulphuric acid?

17. Name the precipitates and their distinctive colours that form in each of the following reactions:
    17.1 Ammonium bromide and silver nitrate
    17.2 Sodium sulphate and calcium chloride
    17.3 Lead nitrate and potassium iodide
    17.4 Sulphuric acid and barium chloride

18. Explain the following conceps:
    18.1 Spectator ions
    18.2 Dissociation of substances
    18.3 Anions
    18.4 Precipitation
    18.5 Halides
    18.6 Electrode
    18.7 Reduction
    18.8 Anode

19. 20 g of lead nitrate, 10 g of potassium chloride and 20 g of potassium bromide are dissolved in water. Half of the lead nitrate solution is added to the potassium bromide and the other half to the potassium chloride.

    19.1 Predict whether any precipitates will form. Motivate your answers with balanced equations.

20. Draw a diagram to represent a Zn-Cu cell under standard conditions.
    20.1 Label the diagram. Indicate the following:

    (a) Anode
    (b) Cathode
    (c) Salt bridge
    (d) External circuit
    (e) Electron flow
    (f) Oxidation and reduction half-reactions
    20.2 What do you understand by standard conditions?

    20.3 Write the cell notation for this cell.

    20.4 Write down the balanced net redox reaction.

ANSWERS

Section A

1. D	2. B	3. D	4.1 A	4.2 B	4.3 C	4.4 B
5. A	6. D	7. C	8. B	9. C	10. A	11. B
12. D	13. A	14. C	15. D	16. B	17. C	18. C
19. A	20. C	21. A	22. C	23. B	24. A	25. D
26. C	27. A					

Section B

1.1 • Sodium chloride solution
   • Dilute hydrochloric acid
   • Tap water

1.2 • Purified water
   • Sugar
   • Copper sulphate crystals
   • Solution of naphthalene in alcohol

1.3 • Dilute hydrochloric acid
   • Sodium choride solution
   • Tap water

2. A solution or molten substance conducting an electrical current is called an electrolyte.

3. The solution of copper sulphate conduct a current because the Cu^{2+} ions and SO_4^{2-} ions act as current carriers. The solution of sugar in water does not conduct an electric current.

4. Different ions in tap water can act as currrent carriers for example the $C\ell^-$ ions in tap water. Purified water is 100% pure and no ions exist to act as charge carriers.

5. Ions

6. False. $PbBr_2$ is not soluble in water. Molten $PbBr_2$ however conducts an electrical current. The Pb^{2+} and Br^- ions act as charge carriers.

7.1 Na_2CO_3

7.2 $Mg(OH)_2$

7.3 $NH_4C\ell$

7.4 $Cu(NO_3)_2$

8. Oxidation – the loss of electrons
   – the reducing agent undergoes oxidation

   Reduction – the gain of electrons
   – the oxidising agent undergoes reduction

9. The voltaic cell: Chemical energy converted into electrical energy.

   Electrolytic cells: Electrical energy converted into chemical energy.

10. Endothermic reaction – is a reaction to which energy must be added continually in order for the reaction to occur.

Exothermic reaction – is a reaction where energy is released so that the chemical potential energy of the reaction mixture is lowered.

11. Electroplating: The covering of metals or other conducting materials with a thin metallic layer.

Preparation of chemicals: Metals, e.g. aluminium and sodium are obtained electrolytically from molten ores.

12.1 Dissociation: $CuC\ell_2 \xrightarrow{\text{water}} Cu^{2+}$ (aq) $+ 2C\ell^-$ (aq)

12.2 Cathode

12.3 Cu^{2+}(aq) $+ 2e^- \rightarrow Cu(s)$

12.4 (a) Cu^{2+}

(b) $C\ell^-$

13.1 $AgNO_3$

$$M_{AgNO_3} = 107,9 + 14 + 3(16)$$
$$= 169,9 \text{ g.mol}^{-1}$$

13.2 $KBr + AgNO_3 \rightarrow KNO_3 + AgBr\downarrow$

$$M_{AgBr} = 107,9 + 79,7$$
$$= 187,8 \text{ g.mol}^{-1}$$

From the reaction follows that:

169,9 g $AgNO_3$ form 187,8 g AgBr

$\therefore$ 4,25 g $AgNO_3$ form $\dfrac{4,25}{169,9} \times 187,8$

$$= 4,695 \text{ g AgBr}$$

14. $CuSO_4 \xrightarrow{\text{water}} Cu^{2+}$ (aq) $+ SO_4^{2-}$ (aq)

$CaC\ell_2 \xrightarrow{\text{water}} Ca^{2+}$ (aq) $+ 2C\ell^-$ (aq)

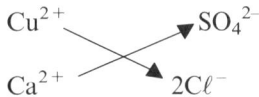

CaSO$_4$ and CaCℓ_2 can form
CuCℓ_2 is soluble
CaSO$_4$ is insoluble; a white precipitate forms

$CuSO_4 + CaC\ell_2 \rightarrow CuC\ell_2 + CaSO_4\downarrow$
white

15.1 $AgNO_3 : AgC\ell\downarrow$
white

15.2 $AgNO_3$ or $Pb(NO_3)_2$: $AgBr\downarrow$ $PbBr_2\downarrow$
light yellow   yellow

15.3 $AgNO_3 : AgI\downarrow$
yellow

15.4 $BaC\ell_2$ or $CaC\ell_2$: $BaSO_4\downarrow$ or $CaSO_4\downarrow$
　　　　　　　 white　　　 white

15.5 • Add a dilute acid to the salt.
　　　• The acid will liberate CO_2 from carbonates.

16. • Add $AgNO_3$ to two different test tubes with samples of both acids

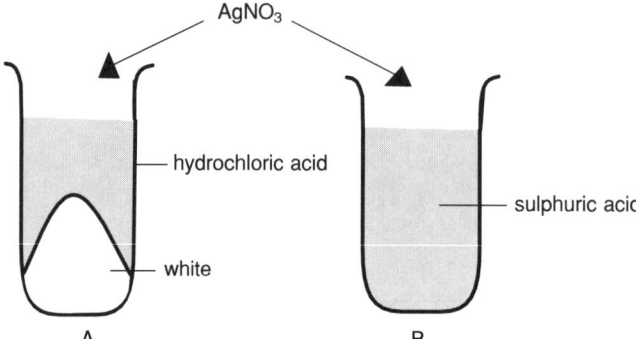

In either A or B a white precipitate forms. That sample is hydrochloric acid.

$HC\ell + AgNO_3 \rightarrow HNO_3 + AgC\ell\downarrow$
　　　　　　　　　　　　　　 white

$H_2SO_4 + AgNO_3$ – no precipitate

• Add $BaC\ell_2$ to two different test tubes with samples of both acids

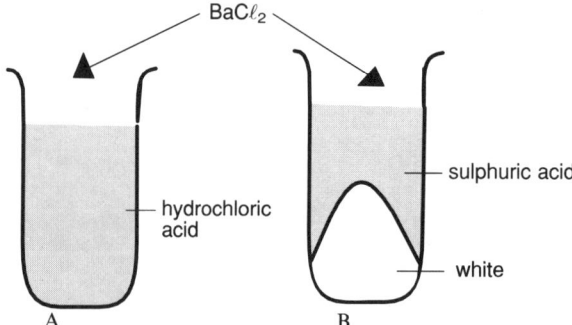

In either A or B a white precipitate forms. That sample is sulphuric acid.
$H_2SO_4 + BaC\ell_2 \rightarrow 2HC\ell + BaSO_4\downarrow$
　　　　　　　　　　　　　　　 white

$HC\ell + BaC\ell_2$ – no precipitate.

17.1 AgBr　　–　light yellow
17.2 $CaSO_4$　–　white
17.3 PbI_2　　–　yellow
17.4 $BaSO_4$　–　white

18.1 Ions not participating in the reaction to form precipitates. They remain in solution.
18.2 The dissociation in water of a solid substance in its composing ions.
18.3 Negative ions.

18.4 A reaction in which precipitates forms.

18.5 $C\ell^-$, Br^-, I^- and F^-; ions of elements of group 7.

18.6 The rod where oxidation or reduction occurs in electrolytic and voltaic cells.

18.7 Reduction is the gain of electrons.

18.8 The electrode where oxidation takes place.

19. $Pb(NO_3)_2 \rightarrow Pb^{2+}$ (aq) $+ 2NO_3^-$ (aq)
$KC\ell \rightarrow K^+$ (aq) $+ C\ell^-$ (aq)
$KI \rightarrow K^+$ (aq) $+ I^-$ (aq)

19.1 Precipitates in both cases
$Pb(NO_3)_2 + 2KC\ell \rightarrow PbC\ell_2\downarrow + 2KNO_3$
white

$Pb(NO_3)_2 + 2KI \rightarrow PbI_2 \downarrow + 2 KNO_3$
yellow

20.

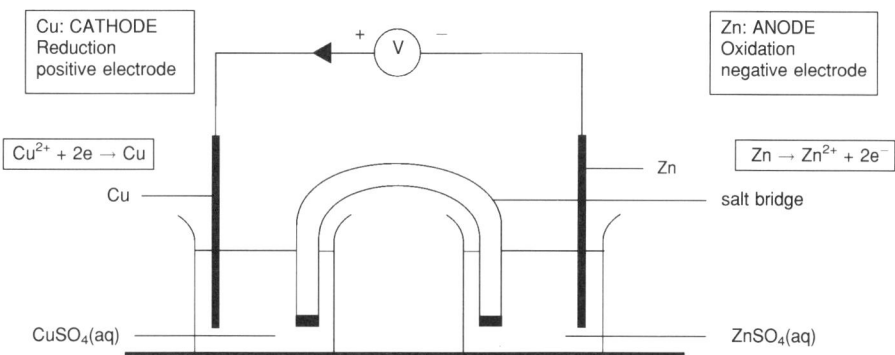

20.7 Standard conditions
- temperature   298 K (25°C)
- Concentration   1 mol.dm^{-3} electrolytes
- pressure   101,3 kPa (where gases are involved)

20.8 $^\ominus Zn|Zn^{2+}$ (1 mol.dm^{-3}) $\|$ Cu^{2+} (1 mol.dm^{-3})$|Cu^\oplus$

20.9 Cu^{2+} (aq) $+ Zn$ (s) $\rightarrow Zn^{2+}$ (aq) $+ Cu$ (s)

12 Temperature, Heat, Work and Change of Phase

1. Temperature

Temperature is a measure of the average kinetic energy of the particles of a substance. An increase in temperature causes an increase in the average kinetic energy of the particles. When the temperature changes, one of the following may also change:

■ Colour

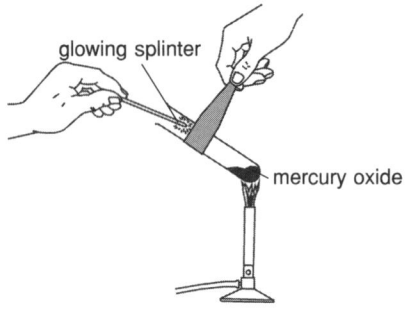

Fig. 1   HgO changes colour during heating.

■ Volume

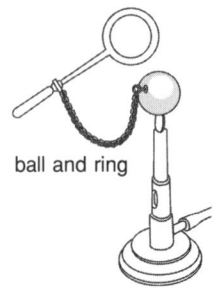

Fig. 2   The volume changes during heating.

An **increase in the temperature of a gas** causes

- an increase in average kinetic energy.
- the particles to move faster and collide more often and with greater impact.
- increase in pressure, because the number of collisions per second determine the pressure exerted by a gas.

■ Phase of the substance

According to the particle model of matter
- all matter consists of small particles
- there are spaces between the particles
- there are repelling and attracting forces between the particles
- the particles are moving continuously.

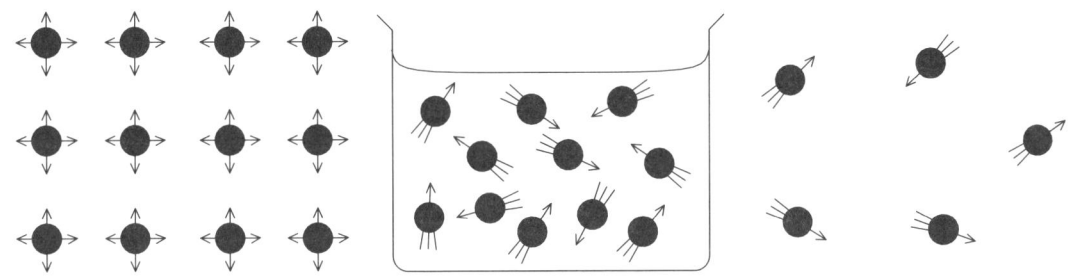

Solids	Liquids	Gases
• Particles are bonded in fixed positions. Strong forces between particles.	Forces between particles are much weaker.	Very weak forces between particles; almost non-existing.
• The particles vibrate constantly in their position of rest.	The spaces between the particles larger, larger volume.	Very large spaces. Move freely.
• Cannot be compressed.	Remains difficult to compress.	Easy to compress.
• —	Exert pressure as a result of impact of the particles on a body.	Exert pressure as a result of the collision of the particles.

- The **Celsius scale is used to measure temperature**. However, a second scale **Kelvin** also exists. In order to convert Celsius temperature to Kelvin, one adds 273.

$$K = {}^{\circ}C + 273$$

- The particles of a substance can have potential and kinetic energy.

2. Heat and Work

- Heat is a form of energy.
- Work and heat are equivalent processes.
- **Heat is energy on its way from a body at a higher temperature to a body at a lower temperature;** from a hot to a cold object.
- Work done on a body or heat transferred to or from a body causes either a change in temperature or a change in phase.

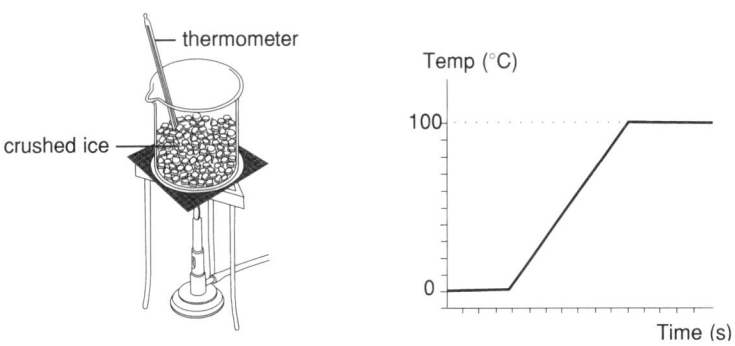

Fig. 3   Phase change during heating.

- The unit of both work and heat is joule (J).
- **Heat** is a term used for **energy transferred from one body to another,** i.e. "energy on its way".

- When equal amounts of heat are transferred to equal masses of the same substance, the rise in temperature is identical.
- When different masses of water are used, the time to cause the same increase in temperature will differ. The more water, the longer it takes.
- For a fixed mass of a given substance, heat transfer is directly proportional to the increase in temperature.
- For a given substance the amount of heat transferred, Q, is directly proportional to the product of the mass and temperature increase.

$$Q \; \alpha \; m \; \Delta T$$

- Different substances require different amounts of heat energy to raise the temperature of equal masses by 1 °C. The **heat capacity (C) of a body is the energy transferred to change its temperature by 1 K or 1 °C.**
- Heat lost by one body is always equal to the amount of heat gained by another body or just heat lost = heat gained.

Example 1:

Calculate the heat energy required to change the temperature of 2 cups (500 ml) of water by 1 °C.

Solution:

According to the definition is the mass of 1 cm^3 (or 1 ml) of water = 1 g
∴ 500 ml water = 500 g water = 0,5 kg
   m = 0,5 kg
    c = 4200 J/kg.K (c is the specific heat capacity of water)

C = m.c
  = 0,5 kg × 4200 J/kg
  = 2100 J

- The **specific heat capacity (c), unit J/K.kg, is the heat energy required to raise the temperature of 1 kg of a substance by 1 K (1 °C).**
  Water has a very high specific heat capacity of 4200 J/kg.K

Specific heat capacities (in Joule per kilogram·Kelvin)

Solids		Liquids	
Copper	390	Alcohol	2 604
Glass	756	Glycerin	2 352
Ice	2 053	Mercury	139
Iron	478	Paraffin	2 688
Lead	130	Petrol	2 142
Zinc	399	Turpentine	1 806

To calculate the energy required to increase the temperature of a substance, use the formula

$$Q = m \, c \, \Delta T$$

Example 2:

The temperature of 3 kg water increases from 20 °C to 25 °C. How much work was done?

Solution:

$\Delta T = 5\,°C$ or 5 K
  m = 3 kg
  c = 4200 J/kg.K

$$Q = m\,c\,\Delta T$$
$$= 3\text{ kg} \times 4200 \text{ J/K.kg} \times 5 \text{ K}$$
$$= 63\,000 \text{ J}$$

Example 3:

Pieces of hot lead, mass 0,9 kg, at a temperature of 71 °C, are added to 680 g of water at a temperature of 20 °C. Calculate the highest final temperature.

Solution:

Suppose the highest temperature reached, is T°C

For lead:		For water:	
	m = 0,9 kg		m = 0,680 kg
	c = 130 J/K.kg		c = 4200 J/K.kg
	$\Delta T = 71\,°C - T$		$\Delta T = T - 20\,°C$

Heat lost by the lead = heat gained by the water.

$$0,9 \text{ kg} \times 130 \text{ J/K.kg} \times (71 - T) \text{ K} = 0,680 \text{ kg} \times 4200 \text{ J/K.kg} \times (T - 20) \text{ K}$$
$$\therefore \quad 8\,307 - 117\,T = 2\,856\,T - 57\,120$$
$$\therefore \quad 2973\,T = 65\,427$$
$$\therefore \quad T = 22$$

$\therefore$ The final temperature is 22 °C.

Example 4:

Pieces of hot metal, mass 0,5 kg, at a temperature of 81 °C, are added to 1 000 g of water, at a temperature of 15 °C. The highest final temperature is 18 °C. Calculate the specific heat capacity of the metal and determine which metal it was.

Solution:

$\Delta T_{metal} = 81\,°C - 18\,°C = 63\,°C$
$\Delta T_{water} = 18\,°C - 15\,°C = 3\,°C$

$Q_{water} = 1 \text{ kg} \times 4200 \text{ J/K.kg} \times 3 \text{ K}$
    $= 12\,600 \text{ J}$

$Q_{metal} = 0,5 \text{ kg} \times c \times 63 \text{ K}$

Heat lost by metal = heat gained by water
$$\therefore \quad 0,5 \times c \times 63 = 1 \times 4200 \times 3$$
$$\therefore \qquad\qquad c = 400 \text{ J/kgK}$$

Metal is zinc. (c = 399 J/K.kg)

Example 5:

400 g of alcohol at a temperature of 40 °C is added to another volume of alcohol at a temperature of 5 °C. The final temperature of the mixture is 12,5 °C. Calculate the mass of the cold alcohol.

$$Q_{\text{hot alcohol}} = Q_{\text{cold alcohol}}$$
$$0,4 \text{ kg} \times 2604 \text{ J/K.kg} \times 27,5 \text{ K} = m \times 2604 \text{ J/K.kg} \times 7,5 \text{ K}$$
$$m_{\text{cold}} = 1,467 \text{ kg}$$

- The specific heat capacity of a substance can be determined by a method of mixtures or with the aid of a joulemeter.

3. Change of Phase and Heat of Transformation

3.1 Evaporation and condensation

Evaporation
- Not all substances evaporate at the same rate.
- Liquids in closed containers only evaporate until the **space above the liquid is saturated with vapour.**
- Saturation of the space above a liquid in a closed container occurs when the rate of evaporation is equal to the rate of condensation at that specific temperature.
- The rate of evaporation also depends on the exposed surface area, temperature, wind and atmospheric pressure.
- **Evaporation** occurs when some molecules have **sufficient kinetic energy** to be able to **overcome the attracting forces of the other liquid molecules**.
- Condensation is the reverse process when vapour molecules lose energy to become liquid again.
- During evaporation the temperature of a substance decreases. This can be explained by means of the kinetic particle model:
  - **All matter consists of moving particles or molecules.**
  - **The particles have kinetic energy.**
  - **Temperature is an indication of the average kinetic energy of the particles.**
  - **Some molecules have more kinetic energy than others.**
  - **When the kinetic energy of some molecules is high enough to overcome the attractive forces of the other liquid molecules, they escape from the liquid.**
  - **Since the molecules with the highest kinetic energy have escaped, the average kinetic energy decreases and the temperature drops.**
- When a liquid boils, evaporation also takes place. Evaporation occurs at any temperature. Liquids boil at specific boiling points.
- The **boiling point** is **that temperature where the vapour pressure is equal to the atmospheric pressure**. It is unique for each substance. Water boils at 100 °C at sea level.

3.2 Energy changes during phase changes

When a substance is heated, one of two things can happen:
- The **temperature rises** while the substance remains in the same phase, because the kinetic energy of the molecules increases.

  OR

- **A phase change occurs,** the temperature and the kinetic energy remain constant.
- The molecules use the energy to move further apart and form a new phase like liquids or gases.
- Work must be done to move the molecules apart and increase the potential energy of the molecules.
- Energy is conserved because the added heat energy increases the potential energy of the molecules.
- The process is called **melting** when a solid becomes a liquid.
- When liquids turn into gases, **evaporation** takes place.

The reverse process can also take place.

- The molecules move closer together.
- The average kinetic energy stays the same but the potential energy decreases.
- The **processes** are **called condensation** or **liquefaction of a gas** and **solidification (freezing) of a liquid** due to cooling.

3.3 Heat of transformation

- **Heat of transformation** is the **energy taken up or given off during the process of a phase change**; it is also known as **latent energy.**
- The **energy required to melt 1 kg of a solid to a liquid without increasing the temperature** is called **specific latent heat of fusion. (L_f:** Unit **J/kg)**
- $Q = L_f.m$ can be used to calculate the energy needed to melt m kilograms of a solid.
- The energy required to vaporise 1 kg of a liquid to a gas without increasing the temperature is called **specific latent heat of vaporisation (L_v:** Unit **J/kg)**
- $Q = L_v.m$ can be used to calculate the energy required to vaporise m kilograms of a liquid.
- **A large amount of energy is required to separate the molecules of a liquid** to form the gas phase, therefore the **latent heat of vaporisation is higher than the latent heat of fusion.**

Example:

Calculate the energy required to change 200 ml ice, temperature $-8\,°C$, to vapour at $110\,°C$ at sea level. (Specific heat capacity of ice is 2053 J/K.kg)

Solution:

Heating the ice from $-8°$ to $0\,°C$:

$Q = m\,c\,\Delta t$
$\quad = 0,2$ kg $\times$ 2053 J/K.kg $\times$ 8 K
$\quad = 3284,8$ J

Melting the ice:

$Q = L_f.m$
$\quad = 336\,000$ J/kg $\times$ 0,2 kg
$\quad = 67\,200$ J

Heating the water from $0\,°C$ to $100\,°C$:

$Q = m\,c\,\Delta t$
$\quad = 0,2$ kg $\times$ 4200 J/K.kg $\times$ 100 K
$\quad = 84\,000$ J

Vaporisation of the water:

$Q = L_v.m$
$\quad = 2\,268\,000$ J/kg $\times$ 0,2 kg
$\quad = 457\,200$ J

Heating the vapour from $100°C$ to $110\,°C$:

$Q = m\,c\,\Delta t$
$\quad = 0,2$ kg $\times$ 4200 J/K.kg $\times$ 10 K
$\quad = 8\,400$ J

Total energy required

$Q = 620\,084,8$ J

QUESTIONS

Section A

I Various possibilities are suggested as answers to the following questions. Choose the correct answer.

1. When a piece of nichrome wire is heated in a flame, it radiates light because . . .
   A energy is transferred from the flame to the wire.
   B nichrome is always luminous.
   C a flame is always luminous.
   D matter consists of very small particles.

2. A rise in the temperature of a gas at constant volume always causes . . .
   A expansion.   C a pressure increase.
   B shrinking.   D colour changes.

3. According to the kinetic particle model, all matter consists of . . .
   A energy.   C molecules.
   B small particles in continuous motion. D crystals.

4. Only in the following case will a transfer of heat take place:
   A From a body at 40 °C to a body at 60 °C
   B From a body at a low temperature to a body at a high temperature.
   C From water at 100 °C to air at 15 °C.
   D When work is done.

5. Which of the following statements is always true?
   A A change in temperature always leads to a phase change.
   B When work is done, a rise in temperature always occurs.
   C The energy that a body contains is called heat.
   D A rise in temperature is always accompanied by an energy transfer.

6. To raise the temperature of 10 kg of water with 10 °C, the following amount of energy is required:
   A 100 N   C $4{,}2 \times 10$ J
   B $4{,}2 \times 10^5$ J   D 100 J

7. Which of the following does not occur when water evaporates?
   A A change in the forces between the water particles.
   B A change in the temperature of the remaining water.
   C A change in the freedom of movement of the water particles.
   D A change in the size of the water molecules.

8. Which of the following curves can be the cooling curve of steam to ice?
   Horizontal axes: Time
   Vertical axis: Temperature

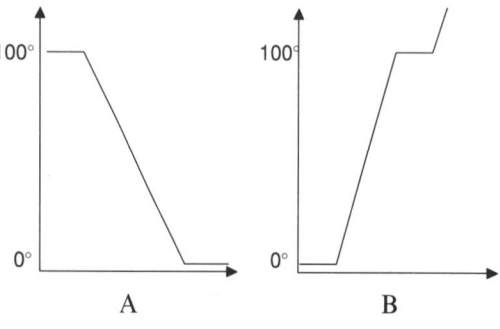
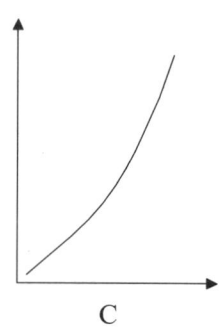
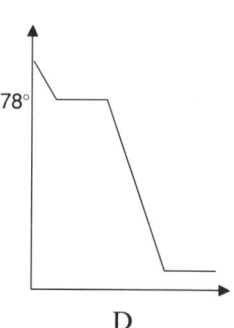

216

9. The energy required to bring about a change of phase without a change in temperature is known as . . .
   A   energy of transformation.
   B   potential kinetic energy.
   C   light energy.
   D   heat energy.

10. When a liquid is heated until it boils, the following change of phase occurs:
   A   Condensation
   B   Solidification
   C   Evaporation
   D   Melting

11. As soon as the **phase change** starts to set in, the additional energy transferred, is mainly used to . . .
   A   evaporate the particles.
   B   cause the particles to expand.
   C   move the particles further apart.
   D   cause the particles to vibrate faster.

12. During **phase changes** an increase in the following type of energy mainly occurs:
   A   Potential
   B   Kinetic
   C   Heat
   D   Motion

13. The phase change that is opposite to evaporation, is called . . .
   A   condensation
   B   boiling
   C   fusion
   D   solidification

14. During spontaneous evaporation, cooling of the remaining liquid also occurs because . . .
   A   the temperature of vapour is very low.
   B   the particles in the gaseous phase possess more energy than the particles in the liquid phase.
   C   the evaporating particles have less energy than the remaining ones.
   D   there are more collisions between the particles of the remaining liquid.

15. Which one of the following curves is the cooling curve where only one phase change sets in?
   Horizontal axis: Time
   Vertical axis: Temperature

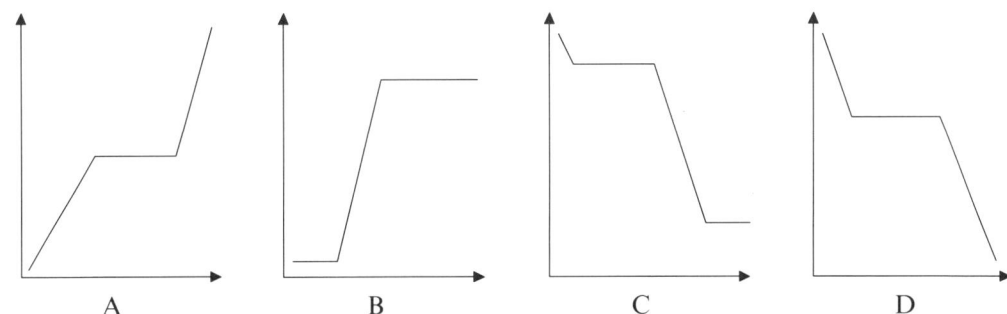

   A   B   C   D

16. The melting point of a substance can be used to identify the substance because . . .
   A   the melting point can be determined by means of a cooling curve.
   B   no two substances have exactly the same melting point.
   C   ice and paraffin have different melting points.
   D   all substances have the same melting point.

17. A change in temperature is an indication of a change in the . . . of the particles of a substance.
   A   potential energy
   B   kinetic energy
   C   internal energy
   D   heat energy

18. The change of sulphur from a solid to a liquid when heated, is probably indicative of . . .
   A a drop in temperature.
   B a decrease in the forces between the sulphur molecules.
   C an increase in the forces between the sulphur molecules.
   D a smaller distance between the sulphur molecules in the liquid phase than in the solid phase.

19. If the volume of a gas is kept constant, a transfer of energy to the gas will lead to . . .
   A a rise in temperature and an increase in the pressure of the gas.
   B a rise in temperature only.
   C expansion.
   D an increase in pressure only.

20. As long as a sample of matter remains in the same phase, a transfer of energy to the sample results in . . .
   A a rise in temperature only.
   B mainly an increase in the potential energy of the particles.
   C a decrease in the kinetic energy of the particles.
   D a decrease in the potential energy and an increase in the kinetic energy of the particles.

21. The SI unit for heat amount is the . . .
   A kelvin (K).   C joule (J).
   B degree Celsius (°C).   D joule per kilogram (J/kg).

22. The SI unit for work is the . . .
   A kelvin (K).   C joule per kilogram (J/kg).
   B newton meter (Nm).   D joule per kilogram per degree Celsius (J/kg °C).

23. The SI unit for energy is the . . .
   A degree Celsius (°C).   C joule per kilogram per degree Celsius (J/kg °C).
   B joule (J).   D joule per kilogram (J/kg).

24. The SI unit for specific heat capacity is the . . .
   A kelvin (K).   C joule per kilogram (J/kg).
   B joule (J).   D joule per kilogram per degree Celcius (J/kg °C).

25. Which of the following is **not** a phase state of a substance?
   A A solid   C A liquid
   B A gas   D A metal

26. When a copper object is heated, . . .
   A the temperature increases but the density decreases.
   B the temperature drops but the density increases.
   C the volume decreases.
   D both the temperature and density increase.

27. While ice is melting . . .
   A its temperature drops.   C its temperature remains unchanged.
   B its temperature increases.   D it gives off heat.

28. The latent heat of fusion (liquefaction) is the same as the . . .
   A latent heat of solidification.   C latent heat of condensation.
   B latent heat of evaporation.   D specific heat capacity of a substance.

29. Consider the following two statements:

I   The amount of heat released when a material solidifies, is equal to the amount of heat needed to melt it.

II   The solidification temperature can always be determined accurately and is the same as the melting point.

A   I is true and II is false   C   Both I and II are false
B   I is false and II is true   D   Both I and II are true

30. Consider the following two statements:

I   During the melting process, the temperature does not change even though heat is still being added.

II   Each substance has its own melting point.

A   I is true and II is false   C   Both I and II are false
B   I is false and II is true   D   Both I and II are true

31. As a block of ice at $0\,°C$ changes into water at $0\,°C$, . . .

A   it increases in volume.   C   its volume stays the same.
B   it gives off heat.   D   its volume decreases.

32. The change of state of a substance directly from a solid to a gas or vapour, is called . . .

A   condensation.   C   solidification.
B   sublimation.   D   evaporation.

33. The change of state of a substance from a liquid to a gas or vapour is called . . .

A   fusion.   C   condensation.
B   evaporation.   D   solidification.

34. The latent heat of evaporation of a substance is equal to its . . .

A   specific heat capacity.   C   latent heat of solidification.
B   latent heat of fusion.   D   latent heat of condensation.

35. The same mass of alcohol at the same temperature is placed in a test-tube and in a saucer. What will happen?

A   The alcohol in the test-tube will totally evaporate before all the alcohol in the saucer evaporates.

B   The alcohol in the saucer will totally evaporate before all the alcohol in the test-tube evaporates.

C   The alcohol in the test-tube and saucer will take the same time to evaporate.

D   The alcohol in the test-tube does not evaporate.

36. 2 g of ether, oil, water and alcohol are dropped alongside each other on the same piece of blotting paper. The order in which they evaporate is . . .

A   alcohol, oil, ether and water. C   ether, alcohol, water and oil.
B   oil, ether, alcohol and water. D   water, oil, alcohol and ether.

Additional for Higher Grade

37. As soon as a substance starts to undergo a phase change, any additional energy transferred to the substance is mainly used to . . .

A   separate the particles from each other and to increase their kinetic energy.

B   decrease the internal energy of the substance.

C   increase the heat content of the substance.

D   move the particles further apart and increase their potential energy.

38. Every minute for five minutes 2,6 kJ energy is transferred to an amount of water. 5 g of water evaporates. The specific latent heat of evaporation is therefore . . .
A   65 kJ/kg.   C   0,065 kJ/kg.
B   2 600 kJ/kg.   D   2 266 kJ/kg.

39. When molten candle wax is dropped on ones hand, it burns more severely when it starts to solidify because . . .
A   solid candle wax is at a higher temperature than the molten wax.
B   molten candle wax is at a higher temperature than the solidified wax.
C   molten candle wax liberates much energy when it solidifies.
D   candle wax has a low latent heat of fusion.

40. The drop in temperature of the remaining water when water evaporates spontaneously, can be ascribed to . . .
A   the change of phase.
B   the escaping molecules with a higher average kinetic energy than those remaining.
C   the remaining liquid that possesses more internal energy than the vapour
D   the higher kinetic energy of the remaining molecules.

41. The excess space in a closed bottle half filled with alcohol is saturated with alcohol vapour. This stage is reached when . . .
A   no further alcohol evaporates.
B   condensation and evaporation occur at the same rate.
C   no more liquefaction of the vapour occurs.
D   the excess space in the bottle occupies exactly half the volume of the bottle.

42. The specific latent heat of fusion of ice is 333 kJ/kg. The amount of energy needed to change exactly 200 g of ice into the liquid phase, is . . .
A   333 kJ.   C   66,6 kJ.
B   1 665 kJ.   D   6,66 kJ.

43. The specific latent heat of evaporation of water is much higher than the specific latent heat of fusion of ice because . . .
A   the temperature of steam is always 100 °C.
B   the average kinetic energy of the ice particles is much greater than that of the steam particles.
C   most of the energy is needed to separate the particles from one another when a change to the gas phase occurs.
D   the molecules in the solid phase is much further apart than in the gas phase.

44. The amount of energy transferred (ΔE), depends on the mass (m) of a substance, the change in temperature reached (Δt) and the specific heat capacity (c) of the substance. Which one of the following equations is correct?
A $m \times c = \Delta E \times \Delta t$ C $\Delta E = m \times c \times \Delta t$
B $\Delta E \times c = m \times \Delta t$ D $m = \Delta E \times c \times \Delta t$

45. 5 litres (5 kg) of water have to be heated through 40 °C. The specific heat capacity of water is 4,2 kJ/kg °C. The amount of heat required is . . .
A   47,6 kJ.   C   200 kJ.
B   168 kJ.   D   840 kJ.

46. The temperature of 4 kg of petrol decreases by 50 °C. The specific heat capacity of petrol is 2,14 kJ/kg °C. The amount of heat released is . . .
A   8,56 kJ.   C   200 kJ.
B   107 kJ.   D   428 kJ.

47. 4 kg of water at 60 °C is mixed with 6 kg of water at 10 °C. The temperature of the mixture will then be . . .

A   25 °C

B   28 °C

C   30 °C

D   35 °C

48. A piece of steel of mass 2 kg must be heated from –15 °C to 35 °C. The specific heat capacity of steel is 0,4 kJ/kg °C. The amount of heat needed is . . .

A   12 kJ.

B   16 kJ.

C   28 kJ.

D   40 kJ.

49. A tank contains 20 litres of oil at 80 °C. The specific heat capacity of oil is 2,7 kJ/kg °C and its density is 0,9 kg/ℓ. The oil must be cooled down to 30 °C. The amount of heat that has to be removed, is . . .

A   1080 kJ.

B   2430 kJ.

C   3278 kJ.

D   3632 kJ.

50. To convert degrees Celsius into Kelvin, we use the formula . . .

A   K = °C + 273.

B   K = °C − 273.

C   K = °C + 100.

D   K = °C − 100.

51. A piece of glass of mass 950 g was cooled from 65 °C to −35 °C. The specific heat capacity of glass is 0,8 kJ/kg °C. The cooling took place at a rate of 3800 J/min. The time taken to complete the cooling process was . . . minutes.

A   6

B   7,5

C   20

D   25

52. The specific latent heat of fusion of aluminium is 322 kJ/kg. This means that it takes 322 kJ of heat energy to . . .

A   raise the temperature of 1 kg of aluminium through 1 °C.

B   cause 1 kg of aluminium to melt at its melting point.

C   cause 1 kg of molten aluminium to evaporate.

D   raise the temperature of 1 kg of aluminium to its melting point.

53. The melting point of ether is –120 °C. The specific latent heat of fusion of ether is 112 kJ/kg. The amount of heat required to melt 4 kg ether is . . .

A   (112 ÷ 4), i.e. 28 kJ.

B   (120 ÷ 4), i.e. 30 kJ.

C   (4 × 112), i.e. 448 kJ.

D   (4 × 120), i.e. 480 kJ.

54. Tin melts at 232 °C and its specific latent heat of fusion is 58,5 kJ/kg. The heat required to melt 1 kg of tin at 232 °C is therefore . . .

A   58,5 kJ.

B   58,5 kJ/kg.

C   232 kJ.

D   232 kJ./kg.

55. 3 kg of ice at −15 °C is added to 10 litres of water at 40 °C. The final temperature is . . .

A   2,5 °C.

B   10,6 °C.

C   32,8 °C.

D   34,3 °C.

56. Consider the following statements:

I   At a high altitude (e.g. on top of a high mountain), the boiling point of water is lower than 100 °C because the atmospheric pressure there is lower than at sea level.

II   The amount of heat required to convert water to vapour at 100 °C is greater than the amount of heat that is liberated when it recondenses.

A   I is true and II is false

B   I is false and II is true

C   Both I and II are true

D   Both I and II are false

57. The specific heat capacity of water is 4,2 kJ/kg °C and the specific latent heat of evaporation is 2250 kJ/kg. The amount of heat required to convert 4 kg of water at 80 °C to 4 kg of water vapour at 100 °C is . . .

A   16,8 kJ.   C   9000 kJ.

B   336 kJ.   D   9336 kJ.

58. 5 kg of water at 20 °C is converted to vapour at 100 °C. The amount of heat required is . . .

A   420 kJ.   C   225 000 kJ.

B   12 930 kJ.   D   900 000 kJ.

59. Ether boils at 34,6 °C and its specific latent heat of evaporation is 380 kJ/kg. The specific heat capacity is 2,27 kJ/kg °C. To cause 20 g of ether at 20 °C to evaporate, the amount of heat required is . . .

A   0,66284 kJ.   C   8,26284 kJ.

B   7,6 kJ.   D   9,17084 kJ.

60. To cause 1 kg of a metal to evaporate at its boiling point, requires 2600 kJ of heat energy. To cause 20 g of it to evaporate needs . . .

A   52 kJ   C   130 kJ

B   113 kJ   D   130 000 kJ

II Complete each of the following by writing down the missing word or words only:

1. An increase in temperature of a substance may lead to changes in . . ., . . . and . . .

2. The phase in which the largest spaces between the particles of a substance exist, is known as . . .

3. During a phase change the . . . remains constant in spite of the fact that a transfer of . . . occurs.

4. Heat is a form of . . . and it is tranferred from an object at a . . . temperature to an object at a . . . temperature.

III If any of the following statements are incorrect, correct them:

1. The spaces between the particles of a substance in the solid phase is much smaller than the spaces between the particles in the gaseous phase.

2. The shrinking of a metal rod can be explained in terms of the fact that the mass of the particles decreases when the substance is cooled down and also that the particles move faster at a lower temperature.

3. The pressure exerted by a gas in a closed container is caused by the collisions of the moving molecules with one another.

4. The temperature of a substance is a measure of how far the particles of the substance are apart.

5. Some of the molecules of a liquid move faster than the other as a result of more collisions with the container in which the liquid is kept.

6. When a substance cools down, it continues to give off energy to its environment and the temperature keeps on decreasing.

Section B

1. You became acquainted with the kinetic model of matter.

1.1 What do you understand by the "kinetic model" of matter?

1.2 Give brief explanations for each of the following in terms of the kinetic model of matter.
    (1) A copper rod expands when it is heated.
    (2) The pressure of a gas in a closed container increases when it is heated.

2. Energy is transferred to matter.

2.1 Mention two changes that the matter can experience (except changes in volume and colour)

2.2 (1) Say with each of the changes mentioned in 2.1 which energy changes occurred.
    (2) What happened to the internal energy in each case?

3. The internal energy of ice is increased when it is heated.

3.1 Briefly explain how the energy of the ice molecules change and which changes accompany the energy transfer.

3.2 Explain why more energy is required to boil water than to melt ice.

4. A small glass beaker with a small amount of ether stands on a few drops of water on a glass plate. Air is blown through the ether.

4.1 Which changes of phase do the
    (1) ether and the
    (2) water experience?

4.2 How does the beaker feel? Explain the observation in terms of the energy of the ether molecules.

5. Steam is cooled down to ice. Molten paraffin wax is cooled down to room temperature.

5.1 Draw rough sketches to present the cooling curves of the two substances.

5.2 Say which parts of the curves represent phase changes as well as which phase changes occur and what they are called.

6. Name two examples of each of the following:

6.1 When work is done, an increase in temperature occurs.

6.2 An energy transfer does not always lead to a change in temperature.

6.3 When work is done, no change in temperature occurs.

7. You have determined the melting point of ice by means of the cooling curve method.

7.1 Which readings had to be taken during the investigation?

7.2 Draw a rough sketch of the curve obtained and indicate the following on the sketch:

    (1) Labelled axes;
    (2) The part of the curve which points to a change of state;
    (3) The reading that has to be made to determine the melting point;
    (4) The part(s) of the curve which point(s) to a decrease in temperature.

8. Briefly explain the following phenomena (in terms of the kinetic theory and energy considerations):

8.1 Water in a canvas bag cools down to below room temperature.

8.2 Washing dries quickly when a wind blows.

8.3 The temperature of boiling water remains constant.

Additional for Higher Grade

9. A small piece (100 g) of hot copper at 120 °C is brought into contact with a big chunk (1 kg) of copper at 20 °C until the heat transfer is completed.

9.1 Which piece of copper experiences the greatest change in temperature?

9.2 How big is the change in temperature in 9.1 above?

9.3 Which one experiences the bigger change in internal energy?

10. An amount of lead shot with a mass of 1 200 g and a temperature of 85 °C is added to 100 g of water at 20 °C in a small fomolite container. The temperature of the water increases to 35 °C. Calculate:

10.1 the amount of energy transferred by the lead shot;

10.2 the heat capacity of the amount of lead shot;

10.3 the specific heat capacity of lead.

11. A mass of 5 kg water at 10 °C must be heated to 80 °C. Calculate the amount of heat required if it is accepted that:

11.1 there is no loss of heat.

11.2 only 60% of the heat is used to heat the water and the rest is lost.

12. A bath contains 50 kg of water at 4 °C. The temperature of the water must be 37 °C. How much water at 92 °C must be added to accomplish this? Accept that there is no loss of heat.

13. The melting point of zinc is 419 °C. The specific heat capacity of zinc is 0,395 kJ/kg °C. The specific latent heat of fusion of zinc is 117,6 kJ/kg. Calculate the total amount of heat required to increase the temperature of 150 g of zinc from 16 °C to 500 °C.

ANSWERS

Section A

I

1.	A	2.	C	3.	B	4.	C	5.	D	6.	**B**	7.	D
8.	A	9.	A	10.	C	11.	C	12.	A	13.	A	14.	B
15.	C	16.	A.	17.	B	18.	B	19.	A	20.	A	21.	C
22.	B	23.	B	24.	D	25.	D	26.	A	27.	C	28.	A
29.	D	30.	D	31.	D	32.	B	33.	C	34.	D	35.	B
36.	C												

Additional for HG

37. D

38. B

39. C

40. B

41. B

42. C

43. D

44. C

45. D

46. D

47. B $\quad 4(60 - x) = 6(10 + x) \therefore x = 18\,°C.$ Final temp. $= 28\,°C$

48. D

49. B $\quad m\,c\,\Delta t = 20(0,9) \times 50 \times 2,7 \text{ kJ} = 3430 \text{ kJ}$

50. A

51. C

52. B

53. C

54. A

55. B $\quad m.c_{ice}.\Delta t + L_f.m + m.c._w \times x = 10(40 - x)\,4200$
$\quad\quad \therefore 3.15.2053 + 336000.3 + 3.4200.x = 1680000 - 42000x$
$\quad\quad \therefore x = 10,6\,°C$

56. A

57. D $\quad 4.20.4200 + 4.2\,250\,000 = 9\,336 \text{ kJ}$

58. B $5.80.4200 + 5.2\,250\,000 = 12\,930$ kJ

59. C $m.c._{ether}.\Delta t + m.L_{v\cdot ether} = 0,02 \times 2,27 \times 14,6 + 0,02 \times 380$
$$= 8,26284 \text{ kJ}$$

60. A $0,02 \times 2600 = 52$ kJ

II

1. phase, volume, density, colour, etc.

2. gas phase

3. temperature; heat

4. energy; high; low

III

1. Correct

2. The particles vibrate slower at a lower temperature. The average distance of vibration around its rest position is smaller. The atoms of the metal therefore occupy a smaller space.

3. Correct – the vibrating molecules not only collide with each other but also with the walls of the container

4. Temperature is an indication of the speed and energy with which the particles of a substance vibrate

5. Some molecules move faster than others as a result of collisions with other molecules

6. Correct

Section B

1.1 • matter consists of small particles
- which exist in continuous motion
- they attract one another but repel one another when they get too close together
- the spaces that the particles occupy is small in comparison with the spaces between the particles

1.2.1 • the copper particles vibrate faster and further away from their equilibrium position.
- this has the effect that the substance needs more space and therefore expand.

1.2.2 • the gas particles move more energetic and
- collide more frequently with other particles
- more collisions and more energetic motion require more space to move in
- the space however is limited and the pressure therefore increases.

2.1 • Change in temperature
- Change in phase (or state)

2.2.1 First: Increase in the kinetic energy of the particles
Second: Increase in the potential energy of the particles

2.2.2 The internal energy of the substance increases in both cases.

3.1 • The temperature of the molecules increases
  • the molecules vibrate further from their equilibrium position without leaving their positions

3.2 • When ice melts, the molecules acquire more freedom but the distance between them remains almost the same
  • When water boils the molecules not only acquire more freedom of motion but also move much further apart.

4.1 • The ether evaporates, i.e. change from the liquid to the gas phases
  • The water will probably freeze, i.e. change form the liquid to the solid phase

4.2 • The beaker feels cold; evaporation of the ether molecules causes cooling
  • If the beaker cools down sufficiently, the water will freeze

---

5.
5.1 Part BC • condensation of the steam
      • steam changes to water

  Part DE • water freezes
       • solidification takes place
       • water changes to ice

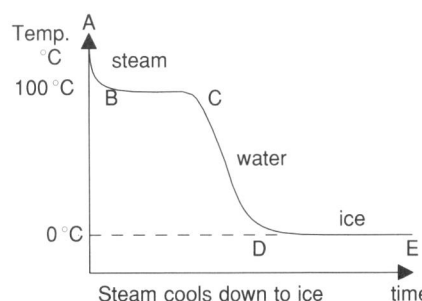

5.2 Part BC • paraffin wax solidifies
      • liquid to a solid

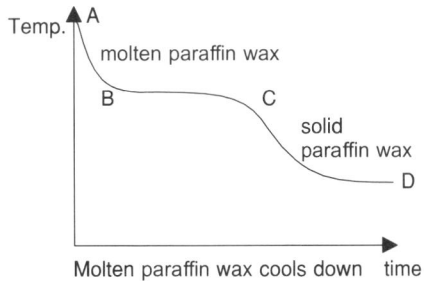

6.1 • Hands are rubbed together
  • An electric drill drills a hole in metal

6.2 • Water at 100 °C changes to steam at 100 °C
  • Ice at 0 °C changes to water at 100 °C

6.3 Same as **6.2**

---

7. • The temperature is read on a thermometer
  • The time is taken on a stop watch or wrist watch

(1)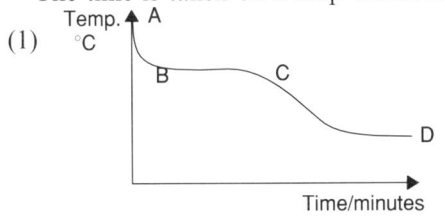

(2) BC indicates a phase change

(Answer 7 continues on the next page)

    (3) The temperature where the graph runs parallel to the time axis is the melting point or solidification temperature

    (4) AB and CD indicate drops in temperature

8.1
- The water seeps through the canvas bag and evaporates
- Evaporation causes cooling of the content of the bag
- The temperature of the water may then drop below room temperature

8.2
- The water of the wet washing evaporates
- When a wind blows, the water vapour is removed and cannot condense again
- In still weather some of the vapour can condense on the washing again

8.3
- When water boils a phase change occurs
- The added heat causes an increase in the potential energy of the molecules
- The kinetic energy of the molecules remains constant
- The temperature of the water therefore remains constant

Additional for Higher Grade

9.1 The small piece of copper

9.2 Heat lost by small piece = heat gained by chunk

$$m.c.\Delta t_1 = M.c.\Delta t_2 \quad \text{(c the same on either side)}$$

$$\therefore \Delta t_1 = \frac{M.\Delta t_2}{m}$$

Because M is 10 times bigger than m, the change in temperature of the chunk is only $\frac{1}{10}$ of that of the small piece. However, we do not know what the final temperature of the two together was.

9.3 Because the internal energy represents the total energy of **all the particles**, the gain in the internal energy of the chunk is exactly equal to the loss of internal energy of the small piece

10.1 Energy transferred by lead shot = energy received by the water

$$= m.c.\Delta t$$
$$= 0{,}1 \text{ kg} . 4200 \text{ J} . 15\,°\text{C}$$
$$= 6300 \text{ J}$$

10.2 Heat capacity $C = \dfrac{Q}{\Delta t}$

$$= \frac{6\,300}{50}$$
$$= 126 \text{ J/°C}$$

10.3 Spes. heat capacity $c = \dfrac{C}{m}$

$$= \frac{126}{1{,}2}$$
$$= 105 \text{ J/kg °C}$$

228

11.1 $Q = m.c.\Delta t$
$= 5 \times 4200 \times 70$
$= 1\,470\,000$ J

11.2 60% of the heat $= 1\,470\,000$ J

$\therefore$ 100% of the heat $= \dfrac{100}{60} \times 1\,470\,000$

$= 2\,450\,000$ J

12. Heat lost by hot water $=$ heat gained by cold water
$\therefore$ m c $\Delta t_1 =$ m c Δt_2
$\therefore$ m . 4200 $\times$ 55 $= 50 \times 4200 \times 33$
$\therefore$ m $= 30$ kg

13. Q $=$ Heat aquired by $\quad +\quad$ heat needed to $\quad +\quad$ heat acquired by
heated zinc $\qquad\qquad$ melt zinc $\qquad\qquad$ molten zinc

$= $ m c Δt_1 + m L_v + m c Δt_2

$= 0{,}15 \times 0{,}395 \times 403 + 0{,}15 \times 117{,}6 + 0{,}15 \times 0{,}395 \times 81$

$= 77{,}63$ kJ heat

(It is assumed that the heat capacity of zinc remains unchanged after it has been melted)